Biofloc Based Nursery Raceway Technology for Vannamei Shrimp

THE AUTHORS

Dr. M. Menaga is currently working as an Assistant Professor in the Department of Aquaculture at Tamil Nadu Dr. J Jayalalithaa Fisheries University and she's also the Student Director of World Aquaculture Society- Asia Pacific Chapter. She has aM.F.Sc in Aquaculture from Central Institute of Fisheries Education, Mumbai and a Ph.D. in Tamil Nadu Dr. J Jayalalithaa Fisheries University (TNJFU). Dr. Menaga specializes in the field of Aquaculture particularly in the development of innovative and sustainable microbial management strategies through understanding of the host-microbe interaction for enhanced microbial stability of advanced culture systems. Among her current research focus is on synergistic interaction of microbes and its pathogenic and probiotic role in aquaculture ecosystem.

She is the author and co-author of various research articles in international and national peer-reviewed journals and books. She has handled the institute industry projects of many international companies and also served a National Organising secretary for the major international event in WAS. International exposure on Aquaculture practices in Indonesia, Taiwan, Malaysia, Myanmar and Singapore were also obtained. She is actively involved in undergoing research in the major areas of aquaculture such as Aquaculture systems, Biofloc technology, functional feed ingredients and aquatic animal health.

Dr. N. K. Chadha is currently heading the Division of Aquaculture at ICAR-Central Institute of Fisheries Education, Mumbai. He has more than 36 years of experience in the field of Fisheries Research, Teaching and Extension. His major area of interests includes Inland Saline Aquaculture, Freshwater Finfis hand Shellfish Culture, Freshwater Finfish and Shellfish Hatchery Seed Production. He was instrumental in establishing Inland Saline Aquaculture Centre of CIFE at Rohtak, Haryana and served as a Founder Office In charge. To his credit, he has published more than 150 research papers in national and international peer reviewed journals and various reports, manuals and books were also published. He has guided students of Masters, M.Phil. & Ph.D. degree. He was also awarded with Best Teacher award and several other awards for his dedicated work and professional skills. He has also acquired international training on Marine Farm System in Japan and also handled several International and National research projects.

Prof. S. Felix has 38 years of teaching and research experience in aquaculture. He is the Past President of World Aquaculture Society-Asian Pacific Chapter and he has also served as Director of the World Aquaculture Society(APC). Prof. S. Felix has served as Second Vice Chancellor of Tamil Nadu Dr. J. Jayalalithaa Fisheries University and also served as Dean of the Fisheries College in Tamilnadu, India. He has developed over a dozen off-campus facilities and 40 Constituent units for the University all over the state for the benefit of students & farmers, besides creating half a dozen new Fisheries Institutes to impart fisheries education. He developed and introduced the raceway culture technology for the first time in India. Under his effective leadership, he has established the Aquatic Rainbow Technology Park and an aquatic Rainbow Mall to the tune of Rs.12 crores for ornamental fish culture as first of its kind in the country. He has 10 patents published on various aquaculture technologies to his credit. He has published 15 Text Books, 50 Manuals and more than 150 Research papers and reports. He has organised 80 Seminars and Conferences. Introduced 30 new courses comprising UG, PG, Ph.D and Paraprofessional programs. Implemented 30 R&D and Technology projects to develop a dozen aquaculture technologies for the benefit of farmers.

PREFACE

Be the change you want to see in the world- Mahatma Gandhi

As per the words of Shri.Mahatma Gandhiji, this book on Nursery Rearing of Vannamei in Biofloc Technology was our unified effort and energy to see the change in the Indian Aquaculture. Biofloc technology allows aquaculture systems to reduce water exchange and usage by preserving adequate water quality within the culture unit while producing low-cost protein-rich bioflocs that may be used as a feed for aquatic creatures. We, the researchers carry out the biofloc experimental trials in the massive culture facility such as indoor raceways, a state of art facility for finding the sustainable solution the aquapreneurs. This book can be used as a starting point because it offers valuable information on how to deploy this technology properly. Despite the fact that the nursery phase of shrimp farming is well established in western countries, the Indian Shrimp Farming Industry is only getting started. Because shrimp culture is still in its infancy, standardization of nursery techniques in various cultures systems, both indoor and outdoor, was successfully achieved. We decided to take a chance in commercial scale systems like raceways and lined ponds after adapting this technique to vannamei nursery rearing on a small scale and seeing its potential in enhanced vannamei shrimp production flashed through our heads as we worked on adapting this technique to vannamei nursery rearing on a small scale. On the other hand, we are attempting to create awareness about the advantages of nursery farming in various systems, particularly in developing nations such as India.

When compared to traditional vannamei growing processes, the use of bio floc technology in shrimp farming for increased production leads to the logical conclusion that this technology will benefit farmers and generate higher returns. It is our sincere hope that this book would aid in the adoption of Vannamei nursery culture procedures in India, hence assisting future shrimp production. We'd also like to thank the Tamil Nadu government for giving significant financing under the Part II scheme, which enabled us to conduct the research on a larger scale, without which this book would not have been feasible!

Authors

CONTENTS

1

CHAPTER

INTRODUCTION

INTRODUCTION

Aquaculture is a major food-producing sector that is growing steadily, coupled with growing population density and land use needs of other industries. Global Aquaculture accounts for about fifty percent of food fish consumption and it is the magnificently rising sector, with an average annual growth rate of about 8 percent. India is second among active countries in aquaculture production with a phenomenal annual growth rate of 4.7 percent, recording 3.2 percent growth in the marine sector and 6.2 percent growth in the inland sector, thereby contributing 1.1 percent to the total GDP and 5.3 percent of the agricultural GDP of the nation. The total fish production in India is 10.1 million tons, producing about 4.6 percent of world fish production and trade to the extent of 2.6 percent in the global fish market. The per capita fish consumption is 9.8 kg, whereas the recommended intake is 13 kg. The projected additional demand for fish in India by 2020 is about 3.21 million t, out of which about 90 percent is expected to be met from the aquaculture sources. The present recognition of aquaculture in India is attributed mainly from carps and shrimps, through foreign exchange earnings. Nevertheless, augmentation of the India's fish production through intensification in a rational way could address the future nutritional food security.

Shrimp farming sector in India is highly remunerative and a fast growing sector with 50240.77 ha area under the culture which yields an export production of 353413.1 MT with an enormous scope for increasing the foreign exchange. Although higher production levels are needed, increased aquaculture production is limited globally by the availability of suitable land and water. The commercial farming of Pacific white shrimp production started from the year 2009-10 is the largest cultured shrimp in terms of production and productivity. Andhra Pradesh tops in the area under culture and production followed by Tamil Nadu and Gujarat.

During the last few decades, the shrimp industry experienced the viral disease out break , events that led to the collapse, or near collapse, of the flourishing shrimp industries in many countries. A disease named firstly as Early Mortality Syndrome (EMS), later defined as acute hepatopancreatic necrosis disease (AHPND), was found in China,2009.And later spread to other countries such as Vietnam, Thailand and Malaysia then spreading to the south and Central America. Though this disease has not been pronounced in India, an equivalent threatening disease known as *Enterocytozoon hepatopenaei* (EHP), *a* microsporidian infects the postlarvae of *Litopenaeus vannamei* has handicapped the Indian Shrimp Aquaculture Industry which leads to the decline in production coupled with a decline in global market prices for shrimp is of serious concern for farmers. The most feasible and environmentally acceptable way to raise aquaculture production is by the use of intensive production systems. Aquacultural practices must be sustainable and minimally destructive to the environment, to maintain the quality and safety standards and enable efficient use of space and natural resources and possibilities for expansion. The water quality control can be done in three different approaches in the intensive production systems

- Replacing pond water with external water, usually at high exchange rates in super-intensive ponds, raceways or tanks.
- Recycling the water through an external bio filter to treat and purify the culture water
- Treating water quality within the pond system using algae (partitioned aquaculture ponds) or bacterial communities (Biofloc/ Aerobic Microbial Floc)

To tackle this, an advance management strategy has been developed to increase juvenile shrimp survival and production predictability for 40-50 days through the advanced systems like raceways and lined ponds for the rearing of post larvae under zero water exchange systems.

Since high water-exchange rates are generally not acceptable due to environmental, biosecurity and water scarcity considerations, other means to control water quality are needed. It has been proved that raceways with limited/ zero water exchange can be used to raise shrimp as a full crop or in a two -phase systems, which helps to minimize the crop loss (Felix.S.,2001). First time in India, Nursery Raceways were used for rearing *Penaeus monodon* in Tamilnadu (Felix.S and Samayakanaan, 2006) .

One of the systems that enable intensification at a relatively reasonable investment and operating cost is Aerobic Microbial Floc (AMF) technology. It is based upon pond management using minimal water exchange and subsequent development of dense microbial populations. The microbes are managed through the adjustment of the carbon: nitrogen (C:N) ratio to control inorganic nitrogen concentrations in the water.

AMF technology takes the advantage of active microbial communities to control water quality and recycle feed. The number of bacteria in zero-exchange intensive ponds can approach 10^9 cells /ml. The bacteria form floc up to a few millimetres in size that are a mixture of bacteria, organic residues and microorganisms such as protozoa and zooplankton.

Proper manipulation of the microbial biomass enables effective water quality control, mostly through conversion of the potentially toxic inorganic nitrogen species to microbial protein in turn, the microbial protein can be utilized to feed the shrimp. The bacteria that form as floc, assimilate total ammonia nitrogen (TAN), produce microbial proteins and enable recycling of unused feed protein.

Vannamei are ideally adapted to AMF system. The filter-feeding omnivores adapt to the harvest of AMF suspended in the water and the strong, stable fish grow and flourish in dense systems. The shrimp biomass can reach 20 mt/ha in well-managed ponds. Shrimp can be grown at very high density in aerated ponds. The practice of zero water exchange in the shrimp culture for the first 50 days is recommended when possible and open water flow through the pond system is discouraged. Shrimp growing in AMF ponds are constantly in contact with a large variety of micro-organisms in the system. The AMF contain a very large variety of organisms, bacteria, algae, protozoa, zooplankton *etc.* It was found, using molecular pyro sequencing technology that a AMF system contains about 2000 different species (Jang *et al.*, 2011).One mechanism toward the control of diseases is the production of short chain fatty acids (SCFA).Adding such acids,(butyric, formic, acetic, propionic or valeric acids) to the culture water by enhancing microbial population resulted in the protection of the culture animal against pathogens. Reducing water exchange can decrease the potential introduction of pathogens into the culture systems and reduce the amount of nutrients being released into the environment without compromising water quality. Zero exchange practices not only reduce the risk of introduction and spread of diseases, it also provides some nutritional benefits from the natural productivity within the system.

The techniques described here are prototyped and currently being used in Fisheries College and Research Institute, Ponneri for the first time in India. An advanced management strategy has been developed to increase juvenile shrimp survival and production for the initial 50 days (the crucial phase of shrimp culture) in tropical Indian conditions. The on-farm trials have been conducted to evaluate the relative efficiency of Aerobic Microbial Floc (AMF) in Indoor raceways and Outdoor lined ponds in the nursery rearing of *Litopenaeus vannamei* for the adoption of AMF driven intensive nursery technology for shrimp farming, which has been proved to be eco-friendly.

- The study was designed and undertaken with the following objectives.
- To compare the efficiency of AMF in raceways and lined ponds at nursery phase of *Litopenaeus vannamei*
- To assess the growth and survival of *Litopenaeus vannamei* in raceways and lined ponds at nursery phase

CHAPTER 2

SUMMARY OF RESEARCH COMPONENTS

Global aquaculture currently accounts for 40% of seafood production and provides 60% of shrimp demand (FAO, 2010). The shrimp aquaculture industry is rapidly expanding and accounts for 15% of the internationally traded seafood products (FAO, 2014).The rapid expansion of this industry has stimulated the intensification of production systems (FAO, 2012). For several decades, the increase in world shrimp production relied mostly on traditional extensive farming practices, but more recently, industry expansion has focused more on intensification. Operating intensive culture systems with consistent production has been a major challenge for shrimp producers. Incorporating a nursery phase into shrimp production systems provides several advantages over direct stocking. It increases shrimp survival, improves facility utilization and provides larger crops per year with better control over predators, water quality, shrimp growth and feed utilization. Furthermore, since post larvae are kept in relatively small enclosed structures, use of nursery facilities in areas affected by viral diseases can serve as a primary quarantine centre, protecting the young shrimp from exposure for a longer period. This expansion resulted in nutrients and organic waste releases, and sometimes in spread of diseases that damaged receiving streams (Cowey and Cho., 1991). Uncontrolled growth already imposed heavy losses and raised major criticisms that threaten the further development of this industry (Naylor *et al.*, 2000).To reduce losses to disease outbreaks, producers have been looking for more sustainable and cost-effective practices.

2.1 NURSERY PHASE INTERVENTION IN SHRIMP FARMING

Nursery phase in shrimp farming have advantages over direct stocking, which increases control over stock inventories, water quality and feed management (Samocha and Benner,2001).It provides for larger and hardier shrimp to be stocked and facilitates a higher number of crops per pond as a shorter period is needed to reach a marketable size (Clifford., 1985; Fast., 1991).The survival of this nursery system is high (85-95 %) with higher anticipated profit .Use of nursery systems can also prevent the spread of diseases as post larvae can be kept under quarantine during the nursery. Thus, a significant improvement in the yields has been reported from white spot infected areas in Ecuador when intensive nursery facilities have been used rather than direct stocking. Farms with intensive nursery facilities can also stock and store PLs in high density to reduce pressure on limited hatchery supplies (Samocha and Benner.,2001).Davis and Arnold.,1998 have reported about the nursery raceway system along with several alterations of procedures and system design to increase the efficiency of waste removal, to minimize labour requirements and to maximize biomass loading. Otoshi *et al.*,2003 used the recirculating raceway systems for the production of brood stock shrimps of *L.vannamei.*

2.2 BENEFITS OF USING SHRIMP NURSERY SYSTEMS

The advantages of two-stage production systems have been summarized by several authors (Fast., 1991; Sandifer and smith., 1977; Samocha and Lawrence., 1992; Felix. S., 2006). In general, two-stage shrimp production systems can improve PL inventory-control and reduce losses to predation. Furthermore, since nursery tanks can be stocked at high densities, they can provide for more efficient utilization of the grow-out space with greater control over water quality and feed consumption. These improved conditions generally result in high PL survival (85-95%) that can lead to greater profit (Hirono., 1983; Aquacop., 1985a; Fast., 1991b; Sturmer *et al.*, 1992; Samocha and Lawrence.,1992). Nursery systems can help to produce larger and hardier shrimp for stocking the grow-out ponds. Furthermore, stocking these larger juveniles can also result in shorter production cycles and a higher number of crops per year (Clifford.,1985; Briggs and Brown.,1991; Fast.,1991, Peterson and Griffith.,1999; Samocha *et al.*, 2001). Because of the shorter production cycle, grow-out ponds should have a better bottom quality that can support higher natural productivity, which can reduce feed cost. Duenas *et al.*,1983 and Hirono.,1989 also mentioned the potential use of nursery facilities for PL holding when grow-out ponds were mostly stocked with wild PL. Use of nursery facilities can also improve the farm performance as shrimp can be kept in the nursery tanks to extend the grow-out ponds dry-out process. Furthermore, in areas known to be severely affected by viral diseases, nursery facilities can be used as primary quarantines to prevent the introduction of diseases into the grow-out ponds. During this waiting period, PL can be inspected and if found to be infected, they can be destroyed with limited effect on the grow-out ponds. Several authors (Samocha and Lawrence.,1992; Peterson and Griffith., 1999; Mc Abee *et al.*, 2003) have suggested the use of greenhouse-enclosed nursery systems in temperate climate areas to stock PL during the early spring season ("headstart"), when ambient water temperature in outdoor ponds is too low for the PL to survive and/or grow. This early stocking can provide for the extension of the grow-out season to facilitate harvesting larger shrimp or for the production of two crops a year (McAbee *et al.*, 2003).

2.3 SHRIMP NURSERIES IN RACEWAYS AND PONDS

Sturmer *et al.*,1992 and Samocha *et al.*,1993a described the first nursery prototype system of six 40 m^3 raceways at the Texas Agri Life Research Mariculture laboratory. Cohen *et al.*,2005 reported good shrimp growth (1.1 g juveniles), high survival rate (>98%), low FCR (<1), and high yields (>4 kg/m^3), while using only low water exchange (1.1%/day). Keeping these shrimp in the nursery system for 50 days had no adverse effects on growth and survival in the grow-out phase at a density of 3,300 PL/m^3.

Harlingen Shrimp Farms, Los Fresnos, Texas, built one of the first intensive nursery-system prototypes in 1990. Good survival and FCR results were obtained when PL was stocked at low (4,000-4,400 PL/m^2) and high densities (11,200 PL/m^2). As documented earlier, the higher stocking density resulted in smaller shrimp than those raised at lower density. The 34-day intensive nursery production in greenhouse-enclosed raceways ,R & G Shrimp, Port Lavaca, Texas reported 100% and 98 % survival ,FCR<1 at a stocking density of 42000 PL/m^2 and 39000 PL/m^2. Although high concentrations of total ammonia-nitrogen and nitrite-nitrogen were found in the culture water of the raceway, no adverse effects were noticed on shrimp health, growth, and survival. Samocha *et al.*,2004 provided a detailed description of the four 97.5 m^2 (147.6 m^3) greenhouse-enclosed raceways that were used for nursery and grow-out of *L. vannamei* using low-salinity ground water (1.8-2.6 ppt). The Stocking and harvest data for postlarvae of *Litopenaeus vannamei* reared in greenhouse-enclosed raceways using low-salinity well water includes 19200PL/m^2 and 20400 PL/m^2 with a 100 % survival on 35 days nursery trial. The data suggested that nursery of *L.vannamei* can be done in an inland facility using low salinity groundwater with good survival, low FCR, and high yields. A follow up grow-out study showed that good survival (86%) and yield (4.39 kg/m^2) were obtained when juveniles harvested from the nursery trial were stocked in the same system. The stocking and harvest data from intensive nursery production trials with the Pacific White Shrimp in a commercial hatchery in Ecuador holds the nursery duration of 8-19 days with 70-99 % survival at a stocking densities 66 PL/L to 83 PL/L.

The Industrias Pecis Shrimp Farm, Mexico used the nursery raceway system mostly to eliminate weak PL before stocking grow-out ponds. In most cases, raceways were stocked at a density of 35,000 PL/m^3 with a typical nursery duration of three to four weeks. The major driving forces of these intensive nursery systems are water circulation, high oxygenation, and suspension of the particulate matter along with careful monitoring of feed consumption. The data presented suggests that good survival, high yields, and low FCR of the Pacific White Shrimp can be expected when using viral-pathogen-free PL under limited water exchange conditions. Certainly, the use of intensive nursery systems can help improve shrimp production practices, especially when combined with better health management, use of high-quality post larvae, and higher quality feed.

Enclosed minimal-exchange systems use considerably less land than that required for open pond culture, can serve as an excellent model for sustainability in aquaculture, and provide an opportunity for inland culture operations (Browdy *et al.*,2001a).Earthen ponds, conventionally used for farming have several disadvantages such as continuous water seepage; complications with soil-water chemistry which have not been understood properly to date; difficulties in determining the causes of water quality related problems; organic load accumulation which cannot be removed during culture; and unavoidable reduction in aeration efficiency and energy loss. Lined ponds, on the other hand, have many advantages. The slightly higher initial investment will be largely compensated for by the huge advantages in terms of net profit. The lining needs to be done with a material that is long lasting and cost effective. A few materials have been tested and found to be effective. Synthetic Fibre Fabric (SFF) Liner Material is a special quality reinforced material proven to be ideal for lining earthen ponds. It is available in various thicknesses viz. 330 to 500 gsm and in different colours (Felix. S *et al.*,2014)

Nursery ponds are usually positioned in farms close to the grow-out ponds to avoid long transfer trips. In many cases, nursery ponds are built adjacent to the grow-out ponds so that the shrimp are transferred without the need for hauling tanks. Thus nursery ponds continue to help assess seed quality and provide better space utilization with more control over the environment. Several authors (Parker *et al.*, 1974; Apud *et al.*, 1979; Duenas *et al.*, 1983; Hirono., 1983; Pretto.,1983; Aquacop., 1984; Issar *et al.*, 1987; Seidman and Issar., 1988; Villalon.,1991), have showed a wide range of pond sizes (0.02-4 ha), stocking densities (28-200 PL/m^2), and nursery durations (25-69 days) being used with variable survival (55-85%) and shrimp size at harvest (0.25-5.3 g). A more extensive review of nursery practices in outdoor ponds was provided by Samocha and Lawrence .,1992.Good growth (1.32 g/wk with 21.2 g average wt. after 106 d) and survival (80%) were reported when the shrimp were stocked (50/m^2) in HDPE-lined outdoor ponds operated with limited water exchange (Cohen *et al.*, 2005). A 16% improvement in survival was reported for ponds stocked with PL from the nursery system. Because of the short nursery cycles, biomass load at harvest, for the most part, was below 2.5 kg/m^3. However, the system was designed to handle a much higher biomass load as demonstrated in several longer nursery trials (50 days) with lower stocking densities (6.3 PL/L), where high yields (4.1 kg/m^3) with good survival (85.7%) and low FCR values (1.1) were reported.

2.4 CONCEPT OF AEROBIC MICROBIAL FLOC TECHNOLOGY

"Aerobic Microbial Floc technology" was coined by Felix.S.,2014. AMF technique uses continuous aeration to enhance the energy yield of microbial feed for the efficient nutrient cycling in the zero or limited water exchange systems. AMF technology finds its importance in the zero water exchange systems, as the utilization of feed by bacteria acting in the presence of oxygen (aerobic conditions) is higher than that of bacteria acting under conditions of low or no presence of oxygen (anoxic or anaerobic conditions). Introducing active aerobic bacterial populations could be an effective strategy in rapid degradation of complex organic compounds and overall improvement of health status of cultured organisms. Bioremediation using selected micro or macro organisms to reduce harmful wastes to less hazardous levels. The principle is that to use selected microorganisms outside the host to create the healthy environment. *Bacillus sp, Vibrio sp, Nitrosomonas sp, Nitrobacter sp, Rhodopseudomonas sp, Aerobacter aerogenosa, Cellulomonas biazotea, Saccharomyces sp,* several algae, and ferns are biological agents recognized for their bioremediation abilities.

Aerobic Microbial Floc Technology reinforces the idea that it is possible to produce aquatic organisms in an intensive and especially bio-safe way using microbial protein recycling under vigorous aeration. The Biofloc technology and Aerobic Microbial Floc Technology falls in the same line whereas the bioremediation through intensive aeration is the main focal point in the AMF technology. Knoesche and Tscheu.,1974 adopted the idea to utilize the heterotrophic bacteria as a natural supplement feed for the growth of shrimp and fishes in aquaculture systems. They used activated sludge to process and treat water in a recirculation system, and proposed to mix produced sludge with grains for later re-use as fish feed for carps. The carbohydrate is added for immobilization of ammonia excreted by the cultured animals (Avnimelech and Lacher.,1979;Boyd.,1985;Muthuwani and Lin.,1995).If carbon and nitrogen are well balanced in the solution, ammonium in addition to organic nitrogenous waste will be converted into bacterial biomass (Schneider *et al.*,2006).Bacterial growth is stimulated by adding carbohydrates and nitrogen uptake through the faeces of cultured animal for the production of microbial proteins (Avnimelech,1999).The promotion of nitrogen uptake by bacterial growth decreases the ammonium concentration more rapidly than nitrification (Hargreaves.,2006).Immobilization of ammonium by heterotrophic bacteria occurs much more rapidly because the growth rate and microbial biomass yield per unit substrate of heterotrophs factors are 10 times higher than that of nitrifying bacteria (Hargreaves., 2006).

Wasielesky *et al.*,2006 examined the effect of natural production in a zero exchange suspended microbial floc based super intensive culture system for *L.vannamei*. The study suggested that there is a positive relationship between the growth parameters and the protein content of the feed in this system. Burford *et al.*,2004 studied the contribution of flocculated material to shrimp, *L.vannamei* nutrition in a high-intensity, zero-exchange system. The study suggests that natural biota in the system can contribute substantially to the nutrition of *L.vannamei*. Hargreaves.,2006 studied photosynthetic suspended-growth systems in aquaculture. The study suggests that the nutritional value of bioflocs to aquatic animals is dependent on several factors: food preference, not only the ability to both ingest and digest it but also the density of suspended particle.

2.5 ADVANTAGES OF AMF

Water quality is the key factor for the success of aquaculture crop and that ensure the survival, production and growth rate of the cultured animals (Boyd., 1990;

Burford., 1997). In a culture technique of AMF technology, the desired water quality is maintained and in situ animal feed is simultaneously produced in the form of biofloc particles (Crab *et al.*,2007).water quality management in the biofloc system is based upon developing and controlling dense heterotrophic bacteria within the culture component (Avnimelech.,2007).The addition of fertilizers or carbon sources directly to the pond increases the natural productivity (Crab *et al.*,2007,Uddin *et al.*,2007)

. Biofloc technology also has potential to protect the cultured organisms from infections with pathogenic bacteria which are responsible for spreading of diseases causing major economic losses in aquaculture (Crab *et al.*,2010).BFT can be a novel strategy for disease management on a long-term basis, in contrast to conventional approaches such as antibiotic, probiotic and prebiotic application (Sinha *et al.*,2008). Recent research focused on the value addition of bioflocs, specifically regarding pathogen control, because infectious diseases destroy the aquaculture industry. It was observed that the regular addition of carbon to the culture media is known to select for poly-hydroxyl alkalonate (PHA) accumulating bacteria (Salehizadeh and Von Loosdrecht,2004) such as *Alcaligenes eutroplus, Azotobacter vinelandii, Pseudomonas oleovorans* and others that synthesize PHA granules. Such granules are synthesized when an essential nutrient like nitrogen is limited in the presence of an excess carbon source (Avnimelech.,1999).These PHAs are polymers of β-hydroxyl short chain fatty acids and if degraded in the gut, they could have antibacterial activity similar to short-chain fatty acids (SCFAs) or organic acids.

2.6 HETEROTROPHIC DOMINANCE IN AMF TECHNOLOGY

The three major aerobic processes in aquaculture reported by Ebeling.,(2006) are heterotrophic assimilation, chemoautotrophic nitrification, and photoautotrophic assimilation. Each of these processes has unique benefits for shrimp biofloc systems and management goals should be carefully evaluated. Some examples of microbial management include increasing the carbon : nitrogen ratio (C:N) to favour heterotrophic nitrogen assimilation (Hari *et al.*,2006),adding bacterial supplements to facilitate nitrification (Kuhn and Drahos.,2011),and encouraging photoautotrophic production by removing particles from the water to increase light penetration (Ray *et al.*,2009).Generally, lower density intensive biofloc system that is exposed to consistent light will be dominated by photoautotrophic assimilation. Heterotrophic –favoured systems in which the C:N ratio is elevated through feed formulation, supplemental carbon addition ,or both can be thought of as mid-level density systems.

Heterotrophic shrimp systems perform well at stocking densities between about 150 and 350 shrimp/m^3. On the high end of the scale are the chemoautotrophic –favoured systems in which nitrifying bacteria convert ammonia to nitrite, and then convert nitrite to nitrate. Super-intensive levels of production, stocked at 300 shrimp/m^3 or more can be achieved by relying primarily on the heterotrophic or chemoautotrophic processes. Heterotrophic bacteria are generally considered to be organic sludge degraders. Heterotrophic bacteria can be either gram-positive (*eg: Bacillus*) or gram-negative (*eg: Pseudomonas*). Some are strictly aerobic, but many are facultative anaerobes (they can survive in both the presence or absence of oxygen). Many species tolerate a wide range of environmental conditions: temperature, pH, salinity, *etc*.

Ponds with the stocking density of 1800-2500 PL/m^2 white shrimp (*Litopenaeus vannamei*) in BFT technology showed the heterotrophic dominance than the chemoautotrophs. It was found that when stocking rates falls below 100 PL/m^2, it became more difficult to establish the heterotrophic community because there were not enough organic substrates to support the microbial growth. They will do completely different functions depending on the level of dissolved oxygen present. Heterotrophic "nitrifiers" can also operate in the reverse direction; that is they can convert nitrate to nitrite or ammonia, especially during times of low dissolved oxygen levels. In a pond, this could potentially happen during the hours before sunrise when DO levels are at their lowest.

Heterotrophic "nitrifiers" prefer to obtain their nitrogen from organic sources such as decomposing organic debris. Heterotrophic "nitrifiers" generally cannot utilize nitrites. Only a few species are capable of reducing nitrite to free nitrogen, but, under strictly anaerobic conditions. Scientific studies indicate that, depending on species, between one thousand to one million heterotrophic bacteria cells are required to perform the same ammonia conversion rate as one Nitrosomonas bacteria cell. Heterotrophic "nitrifiers" can also operate in the reverse direction; that is they can convert nitrate to nitrite or ammonia, especially during times of low dissolved oxygen levels. In a pond, this could potentially happen during the hours before sunrise when DO levels are at their lowest.

Nitrifiers are very slow growing because of the manner in which they must obtain energy. Under optimal growth conditions, they will double in population every 15-24 hrs Heterotrophic bacteria, on the other hand, can reproduce in as little as 15 minutes to 1 hour. Mineralization is accomplished by any of a number of species of heterotrophic bacteria. Species from the genus *Bacillus* are the most common. Some species of heterotrophic bacteria can oxidize or reduce nitrogenous compounds directly to nitrites (NO_2), nitrate (NO_3), or other forms of nitrogen (as NO or N_2). In the absence of an organic nitrogen source, many heterotrophs can utilize ammonia instead. Experimental data has shown that up to one million times more (103 - 106) of these heterotrophic 'nitrifers' are required to perform a comparable level of ammonia conversion that is attained by true autotrophic nitrifiers.

2.7 FACTORS INFLUENCING AMF FORMATION

There are several factors that influence floc formation and floc structure (Schryver *et al.*,2008). Ritvo *et al.*,2003 studied the effect of salinity and the pH on the colloidal properties of suspended particles in super intensive aquaculture systems. The mixing intensity imposed by a selected aeration device at a certain power input will determine the steady-state of the floc size.It is an equilibrium between the rate of aggregation and the rate of breakage and floc size distribution (Spicer and Pratsinis,1996; Chaignon *et al.*, 2002). Dissolved oxygen also influences the biofloc formation. Previous studies showed that biofloc produced at lower dissolved oxygen levels have a higher floc volume index (FVI). The organic carbon sources of choice, to a large extent determine the composition of the floc produced (Mikkelsen *et al.*,1996; Hollender *et al.*,2002; Oehmen *et al.*, 2004).

Organic loading rate,temperature, and pH are the other factors which also governs the biofloc formation (Schryver *et al.*,2008). pH has a significant role in the floc formation both quantitatively and qualitatively in the culture of *P. monodon*, pH of 7.5 is found to be optimum for better flocculation in shrimp aquaculture (Kurup and Devi.,2010).

2.8 AERATION AND OXYGENATION IN AMF TECHNOLOGY

Water movement and oxygenation are imperative issues in super-intensive biofloc shrimp culture systems. Water movement is necessary to keep particles in suspension. If the biofloc particles settle, pockets of anaerobic sludge often form, releasing ammonia or sulfides; sulfide can be extremely toxic to animals and can inhibit nitrification, thereby causing a further increase of toxic nitrogen compounds (Avnimelech and Ritvo.,2003).

Oxygen is a limiting factor in aquatic systems. Solubility in water is very low ,about 8 mg/l in fresh water, yet less, about 6 mg/l in saline water (Colt.,2012).Oxygen limitations lead to reduced production, stress, disease and, in some cases, mortality. Shrimps are generally more sensitive to low oxygen .The oxygen concentration in saline water is lower than that of freshwater (Boyd and Tucker.,1998;Fast and Boyd.,1987).In addition, oxygen saturation is lowered with the increase in temperature. Some of the non-utilized organic carbon is rather stable and is respired way after the feed addition. Another neglected process is the sedimentation of organic residues to the pond bottom, where it undergoes a slow degradation, often along future cycles of culture.

Aeration is a basic means of management and control in BFT ponds. A proper aeration system is essential in order to supply oxygen, to mix the water and to control bottom sludge accumulation. Oxygen added by algae is important during the daytime, yet is of limited significance and reliability in intensive systems. Low oxygen conditions are common in pond bottoms. The accumulation of sludge and the development of anaerobic conditions in the sediment were shown to be a factor limiting fish growth (Avnimelech and Zohar.,1986).The effect of sludge on shrimp growth may be especially hazardous since shrimp live and eat at the pond bottom. Drainage and sludge collection are important parts of pond operation. Drainage of ponds before and during harvest should be done quickly. There is a need for an effective hydraulic gradient and proper pipes/canals to get the water out without obstruction. Ensuring proper slope and water discharge capacity of the drainage canals toward the final outlet should be included in the planning process. It has to be emphasized that for efficient raceway operation, drainage has to be completed in 1-2 days at most and all the water in the raceway has to be drained, without leaving any remnants such as wet puddles.

Aerators are operated to provide other services in addition to oxygen replenishment. It was shown that organic degradation and nitrogen transformation in mixed shrimp ponds are more effective than in stagnant water (Avnimelech *et al.*,1992)

Mixing prevents water stratification ,*i.e.* development of an oxygen-rich, warm layer at the surface of the pond and an oxygen poor layer at the bottom. Another important service of water mixing is the control, or at least the partial control, of sludge accumulation at the bottom of the pond. Oxygen concentration was uniform throughout the water column in the mixed peripheral region, yet clear oxygen stratification was obvious in the central stagnant zone: Oxygen was lower at the bottom during early morning hours as well as during the afternoon hours, when the surface layer was enriched to a level of super-saturation with oxygen.

2.9 CARBON : NITROGEN RATIO FOR MICROBIAL MANAGEMENT

Avnimelech.,1999 revealed that the relationship of the addition of carbon source, reduction of ammonia and production of microbial proteins depends on the microbial conversion coefficient (C/N ratio).He found that the addition of carbonaceous substrate was effective in reducing toxic nitrogen (Ammonia-N and Nitrite-N) in shrimp and tilapia culture. Lancelot and Billen., 1985 observed that when the C/N ratio of the organic matter is higher than 10,immobilization of inorganic nitrogen takes place. Boyd .,1996 reported that if the organic matter is low in nitrogen content (i.e. High C/N ratio).Some of the N for the microbial growth must be obtained from the aquatic environment and will become immobilized as microbial protein(Hargreaves.,2006). Asaduzzaman *et al.*,2008 investigated the effect of C/N ratio in ponds with and without substrate addition for periphyton development on the production of *Macrobrachium rosenbergii*. Increased C/N ratio and substrate improve the heterotrophic bacterial abundance. Higher production and survival were recorded because of (1) a lower inorganic content in the water column,(2) a higher total heterotrophic bacterial (THB) abundance supplying additional single cell protein to augment the prawn production, and (3) an improved periphyton productivity and quality. A concentration of about 10 mg NH_4-N L^{-1} could almost completely be removed within 5 h after the addition of glucose at C/N ratio 10,and without the accumulation of nitrite and nitrate (Avnimelech.,1999).

The organic carbon can be supplied either as an additional organic carbon source (e.g. glucose, acetate, glycerol) or by changing the feed composition thus increasing its carbon content (Avnimelech.,1999). It is possible theoretically to calculate the amount of organic matter needed for an intensive pond, based on the amount of nitrogen excreted by the aquaculture species. The organic carbon source of choice will determine the composition of the flocs produced (Hollener *et al.*, 2002; Oehmen *et al.*,2004).It was observed the costs of the different organic carbon sources will be a determining choice factor (Salehizadeh and Van loosdrecht,2004).

The road to go for BFT is the use of organic carbon sources that are considered low-value products in other processing units as e.g. glycerol, which is a by product of bio-diesel production(Dube *et al.*,2007).Crab *et al.*,2009 reported the effect of different carbon sources on nutritional value of bioflocs, a feed for *Macrobrachium rosenbergii* post larvae. Glycerol, acetate, and glucose were used a carbon source. Different carbon sources not only stimulate specific microbes, but also influences the microbial composition and community organization of the bioflocs. Besides these characteristics, the type of carbon source also influences the palatability and digestibility of the cultured organisms. The study indicates that the bioflocs produced on glycerol gave the best results. However, Crab .,2010a suggested that further research should focus on the use of cheap, non-conventional agro-industrial residues as carbon source and hence upgrade waste to healthy feed.

2.10 USE OF CARBOHYDRATE ENRICHED DISTILLERY EFFLUENT

In India there are about 600 sugar mills and the sugar production in the country is 15 million tons (MT) annually for which about 15.0 MT of cane is crushed (Kumar *et al.*, 1998). These industries release about 7.5 million tons of molasses as by-product. These molasses are used in the distillery industry as a raw material for the production of alcohol. Today there are 257 distilleries in India with an installed capacity of 1.5 billion litres of alcohol production per annum. In the distillery industry, for every litre of alcohol product, about 15 litres of spent wash is released as wastewater. About eight litres of wastewater is generated for every 1 litre of ethanol production in distilleries and the mother liquor left after the sugar production is the spent wash. It is rich in organic material and characteristically less toxic and easily amenable for microorganisms. They do not contain any toxic metals, but rich in plant nutrients, organic compounds and proteinaceous substances (Ramasamy *et al.*, 2007).Due to its richness in nutrients, its application in agriculture as fertilizer has been reported. As its nitrogen content is more than its total organic carbon, they are enriched with fermented carbon source. So there is a possibility of getting about 10 to 11 billion litres of wastewater (Distillery Spent Wash - DSW) from distillery industries alone. Spent wash is of purely plant origin and contains large quantities of soluble organic matter and plant nutrients. As the Spent wash generated in the distilleries have high potential to produce methane, it is subjected to anaerobic digestion for methane recovery. The efficacy of aquaculture as a tool for the treatment of wastes has been demonstrated at Madras, where the effluent after undergoing the biomethanation process, is fed into fish ponds. With production rates of 50 tonnes per hectare per year, about 6 hectares of land area has been shown to be adequate for treating 100m3 of effluents (Rangaswami.,1987).This has opened avenues for utilizing a variety of agro-based industrial effluents for aquaculture practices, largely based on microbiological processes, in terms of oxidation or nutrient removal through algae and other macrophytes. The spent wash when subjected to bio-methanation process (primary treatment) looses up to 85 % of its BOD and becomes distillery effluent. Spent wash is very rich in organic component and possesses high BOD value. Due to its richness in nutrients, its use for the agricultural purposes is very common.

Upon application of spent wash increase in organic matter, nutrient and soluble carbon has been previously reported (Deverajan and Oblisami .,1995). Use of spent wash for agriculture purposes not only prevent the water pollution but also serve as a good source of fertilizer. The spent wash is useful as a fertilizer for production of crops like wheat, maize, sugarcane and rape seed. As far as the bioremediation part is concern use of distillery spent wash in aquaculture for its simple degradation and ammonia assimilation has not been reported. This synergistic way of utilization of distillery spent wash by the microbes for its effective degradation brings an eco-friendly application in Aerobic Microbial Flocculent technology.

2.11 Selection of Species

Not all species are candidates to BFT in aquaculture. Some characteristics seems to be necessary to achieve a better growth performance such as resistance to high density, tolerance to intermediate levels of dissolved oxygen (3-6 mg/l),settling solids in water (~10 with a maximum of 15mL/L of "biofloc volume", measured in Imhoff cones) and N-compounds, omnivorous habits and/or digestive system adaptable to better assimilate the microbial particles(Avnimelech.,2007;Azim and Little.,2008,Taw.,2010).

Litopenaeus vannamei is able to take advantage of the microbial community in the water column. The other benefits include improved growth rate, improved feed conversion ratio, and the contribution of various nutritional qualities by the microbial community (Burford *et al.*, 2004; Moss.,1995; Moss *et al.*, 2006; Wasielesky *et al.*, 2006). Shrimp production enhancement has been attributed to both bacterial and algal (Ju *et al.*, 2009) nutritional components. It forms the major export commodity in the International market. An excessive turbidity may cause negative effects on sensitive shrimp species and not all the species are adaptable to grow in turbid water (Avnimelech., 2006).

BFT is proved to be beneficial for both shrimp and finfish culture (Milstein *et al.*,2001;Burford *et al.*,2003;2004; Serfling *et al.*,2006; Wasielesky *et al.*,2006,Widanarni *et al.*,2012).Later marine shrimps (*L.vannamei, Penaeus semisulcatus, Marsupenaeus japonicas, Farfantepenaeus paulensis*) were also identified as the candidate species for BFT aquaculture (Burford *et al.*,2003,2004;Wasielesky *et al.*,2006; Hari *et al.*,2006; Arnold *et al.*,2009;Kuhn *et al.*,2010;Ballester *et al.*,2010; Megahed,2010; Emerenciano *et al.*,2011;Zhao *et al.*,2010)

2.12 AMF IN SHRIMP AQUACULTURE

The shrimp aquaculture has proven to be destructive to the natural environment and population of aquatic animals (Gowen and Bradbury.,1987,Folke *et al.*,1994;Kautsky *et al.*,1997;Naylor *et al.*,2000;Milewski,2001).Crustacean aquaculture is fraught with environmental problems that arise from (1) the consumption of resources such as land, water, seed and feed (2) their transformation into products valued by society and (3) the subsequent release of wastes into the environment (Kautsky *et al.*,2000;Ronnback.,2001).

The direct impacts include release of nutrient rich water and toxic chemicals, the transfer of diseases and parasites to wild stock. It can also impact the loss of habitat and niche space, and changes in the food webs. In semi intensive and intensive farms artificial feeds provide most of the nitrogen (N),phosphorus (P) and organic matter inputs to the pond system. Only 17 % (by dry weight) of the total amount of feeds applied to the pond is converted into shrimp biomass (Primavera.,1993).The rest is leached or otherwise not consumed, egested as faeces, eliminated as metabolites *etc* leads to the eutrophication. Need was felt to adopt eco-friendly technologies to reduce the adverse impacts of aquaculture on social and physical environments (Alagarswami.,1995).

Intensive aquaculture systems efficiently produce dense biomasses of fish or shrimp and they were facing two major problems. The first is the water quality deterioration caused by the high concentrations of metabolites and the second is the low feed utilization in cases when high water exchange, within or outside the pond system(Avnimelech., 2007).Artificial formulated feed is the major investment in aquaculture and it constitutes 40-60% the of operational cost in shrimp production (Mitra *et al.*, 2005).

The major portion (>80 %) of artificial feed is lost in the aquaculture system as uneaten feed and faeces (Daniels and Boyd.,1989; Siddiqui and Al-Harbi.,1999; Rahman., 2006).The uneaten feed in the system lead to deteriorating the water quality through decomposition (Horner *et al.*,1987; Poxton and Allouse., 1987; Cowey and Walton., 1989, Poxton and Llyod., 1989, Moreira *et al.*,2008). Because of the higher dietary protein in the feed, it deteriorates the water and soil qualities in shrimp grow-out ponds (Boyd.,1989).

The benefits of BFT has been described extensively for shrimp culture (Buford *et al.*,2003,2004; Hari *et al.*,2004,2006; Wasielesky *et al.*,2006) and for finfish culture (Avnimelech.,1997,2007, Milstein *et al.*,2001;Serfling.,2006)

Application of BFT making culture more sustainable by reduces water-based nitrogen species discharged to the environment, reducing the protein content of the feed and increase of total revenue (54%) from the harvested shrimp (Kurup,2009). Enhanced pond productivity through stimulation of suspended algal development and by using them to improved water quality, provide additional food, higher carrying capacity in combination with reduced nutrient discharge, improving environmental sustainability. This technology is simple and cheap, making it also socially and economically sustainable, even for small scale (Hari *et al.*,2004,2006; Varghese.,2007; Kurup.,2009).

BFT has been applied successfully in nursery phase in different shrimp species such as *L. vannamei, P. monodon, F. paulensis, F. brasiliensis*, and *F. setiferus*. The primary advantage observed is related to a better nutrition by continuous consumption of biofloc, which might positively influence grow-out performance a posterior, but was not always the case. In addition, optimization of farm facilities provided by the high stocking densities in BFT nursery phase seems to be an important advantage to achieve profitability in small farms, mainly in cold regions or when farmers are operating indoor facilities. It was observed that presence of bioflocs resulted in increases of 50 % in weight and almost 80 % in final biomass in *F.paulensis* early post larval stage when compared the to conventional clear-water system. This trend was observed even when post larvae were not fed with a commercial feed (biofloc without commercial feed).In *L. vannamei* nursery in BFT conditions, reported survival rates ranging from 55.9 % to 100 % and 97 % and 100 %,respectively. It was demonstrated that the addition of substrates in BFT systems increased growth and further enhanced production, while also contributing to more favourable water quality conditions. According to the same study, growth and survival were not affected by stocking density (2500 vs 5000 PL/m^2),therefore greater production outputs were achieved at the higher density.

Furthermore, it was found that *F. brasiliensis* post larvae grow similarly with or without pelletized feed in biofloc conditions during 30-d of nursery phase, which was 40 % more than conventional clear-water continuous exchange system. Biofloc technology has shown many advantages for nursery-reared ,post-larval Malaysian prawns, including maintenance of optimal water quality, higher survival rate, less expenditure for water and control of waste, higher nutritional quality and smaller impacts on the environment (Perez-Fuentes *et al.*,2013).The BFT was also applied in the nursery rearing of *P. monodon* with an addition of carbon source and artificial substrates. They found that biofloc influences positive growth and production of shrimp juveniles, in addition providing more favourable water quality conditions irrespective of higher stocking densities (Arnold *et al.*,2009). Intensive nursery rearing of *Litopenaeus vannamei* in Aerobic Microbial Floc (AMF) driven raceways offers the tremendous production potential in the low saline raceway system (Felix.S.,2015).

2.13 PROXIMATE COMPOSITION AND FATTY ACID PROFILE OF AMF

According to Akiyama *et al.*,1992, nutritionally floc biomass could provide a complete source of cellular nutrition as well as various bioactive compounds. Burford *et al.*,2004 studied the contribution of flocculated material to shrimp, *L.vannamei* nutrition in a high-intensity, zero-exchange system. The study suggests that natural biota in the system can contribute substantially to the nutrition of *L.vannamei*. Hargreaves.,2006 studied the photosynthetic suspended-growth systems in aquaculture. It also recommends that, the nutritional value of biofloc to aquatic animals is dependent on several factors: food preference, not only the ability to both ingest and digest it but also the density of suspended particle. Wasielesky *et al.*,2006 examined the effect of natural production in a zero exchange suspended microbial floc based super intensive culture system for *L.vannamei*. The study suggested that there is a positive relationship between the growth parameters and the protein content of the feed in this system. Ju *et al.*,2008 studied the amino acid profile of the biofloc and reported that biofloc have better essential amino acid index (0.92-0.93) with histidine and taurine as the most abundant amino acids. However, arginine and lysine were found to be the two most limiting amino acids in the

3

CHAPTER

PROCESS FLOW OF THE RESEARCH

The study was conducted to compare the relative efficacy of Aerobic Microbial Floc (AMF) on water quality, changes in microbial diversity, growth and survival of the Pacific white shrimp (*Litopenaeus vannamei*) fed with MF-CEED (Microbial Floc-Carbohydrate Enriched Distillery Effluent) as a carbon source in the nursery raceways and lined ponds. The sampling was carried out at ten days interval to investigate the growth and survival .

3.1 SITE OF THE EXPERIMENT

The experiment was conducted over a period of 50 days in indoor nursery raceways and outdoor lined ponds at Advanced Research Farm Facility, TNFU-Madhavaram Campus ,an outstation facility of Fisheries College and Research Institute, Ponneri, TNFU.

3.2 EXPERIMENTAL DESIGN

Two nursery raceways of 50m^3 (40 m^3 water holding capacity) and two lined ponds of 100 m^3 (80m^3 water holding capacity) were used for this experiment. The culture systems with the replicates will follow One way ANOVA to find the relative efficiency between the two advanced systems in Indoor raceways and outdoor lined ponds. All the experimental groups were fed with commercially available feed (Growel (Marigold brand) Pvt. *Ltd.*, India) having 30 % CP throughout the trials.

3.3 STATISTICAL ANALYSES

All Statistical analyses were performed using SPSS 16.0 for Windows. Water quality parameters were compared by two-way repeated measures ANOVA with systems as the main factor and sampling time as repeated measures factor. One-way analysis of variance(ANOVA) was performed to examine the difference of growth parameters, proximate composition and nutritional quality of Aerobic Microbial Floc, among the two systems. The differences in the Aerobic Microbial Floc production between raceways and lined ponds were compared by one-way analysis of variance. All mean separations were carried out by Duncan's Multiple Range Test at the significance level of = 0.05.

3.4 BIOSECURITY MEASURES

Biosecurity measures include fencing of lined ponds with 250 µ mesh nets to prevent crustacean carriers from entering the modules, eradicating frog larvae, crab fencing, bird scare lines and water aging in the lined ponds. The ponds were lined with 300 GSM nylon fabric sheet to prevent the erosion of the soil. Strict biosecurity protocols were applied on movements of farm personnel, equipment and guests entering farm facilities.

3.4.1 Configuration of Nursery Raceways

The two raceways were operated with high-pressure rapid sand filters. Every raceway had a central longitudinal partition placed over a 5.1 cm (2 in.) schedule 40 mm dia PVC pipe designed to provide oxygen-rich water across the bottom of the raceway. Water oxygenation in each raceway was generated by a 10 hp and 3 hp blowers. This setup increased dissolved oxygen levels of the water going back into the raceway by mixing it with air or pure oxygen demand of the system. In addition, each raceway was provided with eighteen 5.1 cm airlift pumps arranged in three equidistance banks on each side of the partition and six 1m long air diffusers. The daily water exchange rates were calculated based on the total water used in each raceway over the duration of the trial.

3.4.2 Water source and Treatment

Metahaline seawater (90-95 ppt) collected from the nearby lagoon was stored in 5000L reservoir pond. The seawater was allowed to settle down for a week. Meanwhile, the water will have been aged at least 78 hours within the reservoir ponds.This aging process eliminates the viral particles in the water. The most important factor is to make sure the screening, chemical treatment, and aging process are efficiently used before stocking the ponds and raceways. The water was chlorinated with a commercial grade calcium hypochlorite (Bleaching powder of 33 % of active chlorine) to achieve an initial chlorine concentration of 20ppm with targeted residual chorine of 1ppm after 38 hrs. After dechlorination, with sufficient aeration the salinity was adjusted for 7 ppt with freshwater. The diluted water is directed to the sand filter for further refinement and it was filled in the raceways and lined ponds.

3.4.3 Aerobic Microbial Floc production

The treated water is added with the following fertilizers for the development of AMF by following Nyan Taw *et al.*,2006 protocol with slight modifications to promote heterotrophic condition for the culture. The quantity of fertilizers added for AMF development in indoor raceways and lined ponds were presented in table 1 & 2.

Table-1 Quantity of Fertilizers added for AMF Development in indoor Raceways

Day	Fertilizers	Quantity(g)/ton
1	Urea	1.1
	Triple Super Phosphate (TSP)	0.14
	Grain pellet	4
	Dolomite	7
2	Grain pellet	4
	Dolomite	7
3	Grain pellet	4
	Dolomite	7
4	Grain pellet	4
	MF-CEED	10.1
5	Grain pellet	4
6	MF-CEED	7

Table-2 Quantity of Fertilizers added for AMF Development in Outdoor Lined Ponds

Day	Fertilizers	Quantity(g)/ton
1	Urea	1.1
	Triple Super Phosphate (TSP)	0.14
	Grain pellet	4
	Dolomite	7
3	Grain pellet	4
	Dolomite	7
6	Grain pellet	4
	Dolomite	7
10	Grain pellet	4
	MF-CEED	10.1
15	Grain pellet	4
20	MF-CEED	7

3.5 MF-CEED FORMULATION

Distillery spent wash was collected from M/s. Rajshree Biosolutions situated at Coimbatore district, Tamilnadu, India. The physicochemical characteristics such as pH, TDS, TSS, BOD were analyzed by standardized methods (APHA., 2005) in factory lab. Temperature and pH of the spent wash were measured and kept in a controlled environment.

The growth of the microbial community is encouraged with the addition of carbon sources through fermentation. This leads to the mass proliferation of heterotrophic bacteria which in turn reduce the toxic ammonia levels in a few hours. The addition of carbon to increase the Carbon: Nitrogen ratio is done for achieving the better bioremediation. It perhaps provides an effective means to tackle water quality and disease problems in a more natural way. They would efficiently recycle the waste material by decomposing the complex organic substances and generating nutrients in the form of simple inorganic compounds.

Yeast-based fermentation process using *Saccharomyces cerevisiae,* for ease of handling and process economics still remains the first choice for treating the industrial effluents. And with the locally available carbon source, rice flour has been added @ the ratio of 3:1 *i.e* 3 parts of rice flour to the 1 part of distillery spent wash was added. To enhance the fermentation process, the yeast should be activated by boiling in the water and then it is added to the mixture. About 25grams of yeast is requires for 10 litres of mixture. Sufficient aeration should be provided overnight. The fermented mixture obtained known as Microbial Floc- Carbohydrate Enriched Distillery Effluent (MF-CEED) to increase the microbial floc production is then added to the culture systems.

By adding carbon source, the C:N ratio approximately 15:1 was maintained throughout the experiment to optimize heterotrophic bacterial growth (Avnimelech.,1999).

3.6 CALCULATION OF CARBOHYDRATE REQUIREMENT

The quantity of carbohydrate added to the systems were calculated using the equation suggested by Avnimelech.,1999 with an estimated carbon 26%, the CH addition needed (CH) to reduce the total ammonia nitrogen concentration by 1g m^{-3} is 20 g m^{-3}.

$$CH = N/0.05 \qquad ...(1)$$

It can be assumed that the ammonium flux into water , NH_4^+ directly by excretion or indirectly by microbial degradation of the feed residues, is roughly around 50% of the feed nitrogen (Avnimelech,1999)

$$N = \text{Quantity of feed x \% N in feed x \% N excretion} \qquad ...(2)$$

The amount of carbohydrate addition needed to assimilate the ammonium flux into microbial protein was calculated using equation (1) and (2):

CH= Quantity of feed x % N in feed x % N excretion/0.05 ...(3)

Shrimp were fed with commercial feed pellets containing 30.25 % protein at a daily rate of 1.5 % body weight. It was assumed that 33 % of the feed nitrogen is excreted. According to equation (3), 480 g of carbohydrate is required for each kg of 30.25% protein feed.

3.7 STOCKING OF SHRIMP SEED

The specific pathogen free (SPF) *L.vannamei* seed samples from M/s. Nellore Hatcheries,25/2/1931, NGO Colony, B.V.Nagar, Nellore District ,Andhra Pradesh were brought for testing the various pathogenic infections and they were acclimatized at 7 ppt. The seed samples were tested for WSSV ,AHPND and EHP infection two to three days prior to stocking using two-steps PCR. The samples are given to the State Referral laboratory for Fish disease Diagnosis and Aquatic Animal Health. The screening of WSSV, AHPND and EMS was done in Padhmanaba Labs Pvt *Ltd*, Vettuvankeni, Chennai. The seeds are screened for the following infections and the results are mentioned in Table 3.

Table-3 Screening results of *Litopenaeus vannamei* postlarvae from various diseases

S.No		Raceways	Lined ponds
1	Microscopic		
	a. Stress test survival in Formalin	80%	70%
	Salinity test	80%	70%
	b.Size variation	10%	Above 15%
	c.Necrosis	10%	Above 15% (Heavy)
	d.Muscle gut ratio	3:1	3:1
	e.MBV	Negative	Negative
	f.BP	Negative	Negative
	g.HPV	Negative	Negative
2	PCR		
	a.IHNNV	Negative	Negative
	b.TSV	Negative	Negative
	c.HPV	Negative	Negative
	d.WSSV	Negative	Negative
	e.EMS /AHPND	Negative	Negative
	f.EHP	Negative	Negative

MBV-Monodon Baculo Virus: BP-*Baculovirus penaei*: HPV-Hepatopancreatic parvovirus: IHNNV-Infectious Hypodermal and Haematopoietic Necrosis Virus: TSV-Taura Syndrome Virus: WSSV-White Spot Syndrome Virus: AHPND-Acute Hepatopancreatic Necrosis Disease: EHP- *Enterocytozoon hepatopenaei*

3.8 STOCKING DENSITIES IN RACEWAYS AND LINED PONDS

The PL14 (±0.002g) was stocked in the nursery raceways at a stocking density of 6000PL/m^3 in an AMF developed water with floc volume of 5 ml and TSS was 80 mg L^{-1}.The blowers were operated continuously to maintain the dissolved oxygen content in the saturation level. And in the case of Lined ponds the PL14 (±0.002g) were stocked at a density of 2300PL/ m^3 in the floc water of 3 ml of floc volume and 60 mg L^{-1}.The two Aspirators (2HP) were allowed to run alternatively throughout the experiment in the lined ponds.

Growel manufactured crumble feeds, Marigold brand (0.5-1mm) containing 30 % crude protein (Correia *et al.*,2014) were used throughout the experiment. Proximate composition of the experimental diet and the physico chemical properties of the carbon source are given in table 4 & 5. Feeding rates were based on observation of feeding behaviour of shrimp in the treatments during the first few days and fixed at 2 % of the total stocked biomass and adjusted weekly after every sampling. Daily feed rations were split into six equal amounts given at 6:00,10:00,13:00,18:00,20:00 and 2:00 hrs to all the experimental groups.

Table-4 Proximate composition of experimental feed on dry weight basis (g/100g)

Constituents	% on Dry matter basis
Crude protein	30.25
Ether Extract	6.00
Crude fiber	2.73
Total Ash	10.65
Nitrogen free extract	41.42

Table-5 Physico-chemical properties of Distillery Spentwash

Parameters in mg/L	Results
Colour	Dark Brown Colored Liquid
pH @ 25°C	6.25
Total Chlorides	7729
Total Suspended Solids	42730
Total Phosphorus	66.6

Total Solids	326980
Calcium	6072.1
Magnesium	4421.4
Potassium	23055
Carbonates	Nil
Electrical Conductivity Micro ohms/cm @ 25 degree	112652
Total Organic Carbon	26%
Total Nitrogen	834
Organic Solids	252340
Inorganic Solids	74640

3.9 MEDIA, CHEMICALS AND GLASSWARES

Chemicals and reagents were procured from Ganesh Scientific Suppliers ,Chennai; Dolphin Equipments, Chennai. The bacteriological agar media and agarose were procured from Hi-media. The PCR master mix, DNA molecular weight markers and other molecular biology grade chemicals were procured from Hi-media. The glass wares and plastic wares used in this study were procured from Borosil, Tarson and Merck, Mumbai.

3.9.1 Sterilization

The glassware used in molecular and microbiological work, were sterilized in hot air oven (DTC-96, Expo Everflow, Chennai) at 180°C for two hours. The media, pipette tips *etc* were sterilized by autoclaving at 121°C for 15 minutes.

3.10 PHYSICO-CHEMICAL PARAMETERS OF WATER

Dissolved oxygen (DO) was examined for six times (04:00, 09:00, 13:00, 16:00, 21:00, 01:00) in a day. Temperature, pH and Ammonia-NH_3-N were monitored 3 times at 04:00, 12:00, 18:00 daily. The Nitrite-Nitrogen (NO_2-N),Nitrate-Nitrogen(NO_3-N), Phosphate(PO_4), Total Alkalinity, Calcium Hardness, Magnesium Hardness, Turbidity and Imhoff cone volume were tested at 11.00 am periodically. The Total Suspended Solids (TSS), Volatile Suspended Solids (VSS), Total Dissolved Solids (TDS),Turbidity were measured for every three days. The five-day biochemical oxygen demand (BOD_5) was recorded throughout the experiment. Besides these, sodium and potassium were determined weekly once during the experiment. The Floc Volume (ml L^{-1}) was monitored by using Imhoff cones daily.

3.10.1 Temperature

It is measured by Mercury Thermometer.

3.10.2 pH

The pH meter is calibrated and standardised for measuring pH from the water sample.

3.10.3 Total Alkalinity

Total Alkalinity was measured by titration method (APHA,2005) and its values were expressed in mg/l $CaCO_3$

3.10.4 Calcium Hardness

Calcium was measured by Titration method (APHA,2005) and its values were expressed in mg/l.

3.10.5 Magnesium Hardness

Magnesium was measured by Titration method (APHA,2005) and its values were expressed in mg/l.

3.10.6 Total Hardness

Total Hardness was measured by Titration method (APHA,2005) and its values were expressed in mg/l $CaCO_3$

3.10.7 Dissolved oxygen (DO)

Dissolved Oxygen levels were determined by Winkler's titration method(APHA,2005). Its concentration was expressed in mg/l.

3.10.8 Free Carbon dioxide (CO_2)

Free Carbon dioxide was measured by titration method (APHA,2005) and its values were expressed in mg/l. Both carbonates and bicarbonate alkalinities were measured.

3.10.9 Ammonia (NH3-N)

Ammonia levels were measured using spectrophotometer at a wavelength of 640 nm by phenate method (APHA,2005) and compared with standard graph. Its concentration was expressed in mg/l

3.10.10 Nitrite-N (NO2-N)

Nitrite concentration was measured as per APHA,2005 at 543 nm and compared with a standard graph. The concentration was expressed in mg/l.

3.10.11 Nitrate-N (NO3-N)

Nitrate concentration was measured (APHA.2005) at 543 nm and compared with a standard graph.Its values were expressed in mg/l.

3.10.12 Organic Phosphate (PO4-P)

Organic Phosphate concentration was measured (APHA,2005) at 882 nm and compared with a standard graph. Its values were expressed in mg/l.

3.10.13 Sodium

Sodium levels were measured using Flame photometry at a wavelength of 589 nm (APHA,2005) and compared with a standard graph. Its concentration was expressed in mg/l

3.10.14 Potassium

Potassium levels were measured using Flame photometry at a wavelength of 589 nm (APHA,2005) and compared with standard graph. Its concentration was expressed in mg/l

3.10.15 Salinity

Salinity was estimated by adopting Harvey's direct titration method (APHA,2005) and its values were expressed in ppt.

3.10.16 Total Suspended Solids (TSS)

Total suspended solids are determined by filtering the water sample (normally 10-20ml) through a pre-weighed filter paper (Whatman GFC or similar),evaporating the water in an oven (65°-100° c) for a few hours and re-weighing (APHA,2005) and is expressed in mg/l.

3.10.17 Total Dissolved Solids(TDS)

Total Dissolved solids are determined by using TDS meter by filtering the water sample

3.10.18 Volatile Suspended Solids (VSS)

VSS is obtained by weighing the dry filter following TSS determination in a muffle at 550° C and weighing the residue, corresponding to the ash fraction. The VSS is the organic fraction is the TSS minus the ash.

3.10.19 Turbidity

The turbidity of the water sample is measured by using Nephalometer.

3.10.20 Bio-Chemical Oxygen Demand

Dissolved oxygen was measured by Winkler's method before incubation and after incubation at 20°C for 5 days in closed BOD bottles without seeding. BOD was calculated by subtracting the final DO from initial DO, BOD of the sample is expressed as mg/l

$$BOD_5\,(mg/l) = \{ (\text{Initial DO} - \text{Final DO}) \times VB \} / Vs$$

Where,

VB = Volume of BOD bottle

Vs = Volume of a sample taken for titration.

3.10.21 Floc Volume

Floc concentration was measured by allowing the floc to settle down in the Imhoff cone for 15 minutes without disturbance. The concentrations of settled floc were expressed in ml/l.

3.11 CALCULATION OF WATER EXCHANGE RATE

According to Avnimelech.,1999 the water replacement was done 10 % in raceways and 20% in Outdoor lined ponds (adapted values are adjusted according to field environmental conditions) on every third day. Totally 17 top offs were done in 50 days trial throughout the experiment.

3.12 GROWTH PARAMETERS

The growth parameters of the vannamei were estimated by taking their body weight at every ten days interval. The different growth parameters observed during the experiment are given as follows:-

3.12.1 Average body weight (ABW)

The Average body weight was calculated using the following formula

$$\text{Average body weight (g)} = \frac{\text{Total weight (g)}}{\text{Total number of animals}}$$

3.12.2 Specific growth rate (SGR)

The specific Growth rate was calculated by the following formula

$$\text{SGR} = \frac{\text{Log}_e\text{Final weight} - \text{Log}_e\text{Initial weight}}{\text{Number of days} \times 100} \qquad ...(1)$$

3.12.3 Feed conversion ratio (FCR)

The feed Conversion Ratio was calculated by the following formula

$$\text{FCR} = \frac{\text{Feed given (Dry weight)}}{\text{Body weight gain (Wet weight)}} \qquad ...(2)$$

3.12.4 Protein efficiency ratio (PER)

Protein efficiency ratio was calculated by the following formula

$$\text{PER} = \frac{\text{Body weight gain (Wet weight)}}{\text{Protein feed}}$$

3.12.5 Yield (kg/cubic metre)

Yield was expressed in kg/m^3

Final biomass (kg) /Volume of experimental unit (cubic metre)

3.12.6 Survival (%)

Survival was expressed as the percentage of live shrimp on the final day of the experiment relative to the total initially stocked shrimp.

Survival rate (%)=100 (Nf-Ni)/Ni

Ni is the initial number of Shrimp and Nf is the final number of Shrimp.

3.13 CALCULATION OF FEED ECONOMICS

The potential savings on feed that can be obtained by AMF has been calculated according to Avnimelech,2015. Shrimp juveniles produced with commercial feed at a 30% protein content and at a feed conversion ratio obtained from the experiment.

3.13.1 Calculation of the cost saving by the application of AMF

The cost for the production of 1 kg shrimp live weight with AMF (Avnimelech., 2015)

- FCR x cost of per kg feed = cost of 1 kg shrimp live weight produced in raceways
- FCR x cost of per kg feed =cost of 1 kg shrimp live weight produced in lined ponds

3.13.2.Calculation of the amount of organic carbon needed to grow the flocs

The flocs have a C:N ratio of 4 (Avnimelech,1999)

- 4 × 0.04 =0.16 kg C in floc biomass is produced kg^{-1} shrimp live weight produced since all the excess nitrogen should be assimilated in the floc

The yield of bacterial biomass can be added in the water for the floc to be able to assimilate the excess nitrogen kg^{-1} shrimp live weight produced

MF-CEED (26% C) is used as organic carbon source

- 0.32/0.26 =1.23 kg MF-CEED needs to be dosed to the water kg^{-1} shrimp live weight produced.

3.14 PROXIMATE ANALYSIS OF THE AEROBIC MICROBIAL FLOC

At the end of the experimental trial, the microbial flocs were harvested from each raceways and lined ponds and the same were used for analysis of proximate composition. The AMF were collected and these samples were centrifuged at 6000rpm and allowed to dry in an oven at 42°C to constant weight. Proximate composition analyses of the floc were performed as per standard methods were performed following the AOAC ., 2005 methodology.

3.14.1 Crude Protein (CP)

Nitrogen content of the dried samples of the microbial flocs were estimated quantitatively by Kjeltec semi-automated method (2200 Kjeltec Auto Distillation,Foss tecator,Sweden) using titration as the means for determining nitrogen percentage. The crude protein percentage was obtained by multiplying the nitrogen percentage by a factor of 6.25.

$$\text{Crude protein } (\%) = N_2\ (\%) \times 6.25 \qquad \ldots(6)$$

Where N_2 is total nitrogen

3.14.2 Crude Lipid

By petroleum ether method, the fats are extracted from the sample with petroleum ether Calculations.

$$\text{Crude lipid } (\%) = 100\ \frac{B - A}{C}$$

Where:

A = weight of clean dry flask (g)

B = weight of flask with fat (g)

C = weight of sample (g)and evaluated as a percentage of the weight before the solvent is evaporated.

3.14.3 Crude Fiber (CF)

The crude fiber was estimated by using automatic fiber analysis system (FIBROTRON-B 8,Tulin Equipments, India).The fat and moisture free sample (1.0g) were placed in a pre weighed crucible and fitted into the instrument. The samples were refluxed with 100 ml of 1.25 % sulphuric acid followed by 1.25 % NaoH. Each washing continued for 45 minutes. After completion of both the washes, the crucibles were dried in the hot air oven at 105°C for 3-4 hours. The crucibles were cooled in the desiccators and weighed. The dried samples were ignited at 550°C for 4- 5 hours, and then the crucibles were cooled and weighed. The weight difference in the samples before and after ignition gives the crude fiber of the sample and it is expressed as weight in percentage.

$$\text{Crude Fiber } (\%) = \frac{\text{Wt. sample before ashing} - \text{Wt of sample after ashing} \times 100}{\text{Weight of sample}}$$

3.14.4 Total Ash

Ash content of the microbial floc samples was estimated by taking a known weight of dried samples in a silica crucible and placing it in a muffle furnace at 600 °C for 5-6 hours. The calculation was done as follows:

$$\text{Ash } (\%) = \frac{\text{Weight of ash (g)}}{\text{Weight of sample (g)} \times 100}$$

3.14.5 Ether Extract (EE)

Ether extract of dried microbial floc samples were estimated by Soxhlet apparatus (pelican equipment, India) using diethyl ether (boiling point 55±5 C) as the solvent. The calculation was made as follows:

$$EE\ (\%) = \frac{\text{Weights of ether exxtract (g)}}{\text{Weight of the sample (g)} \times 100}$$

3.14.6 Total Carbohydrate (TC)

The total carbohydrate (TC) of the floc samples were calculated by subtracting the percentage of other nutrients from 100 (Hasting.,1969).

TC (%) = 100 – (Crude protein + Ether extract + Ash)

3.14.7 Gross energy (GE)

Gross energy content of AMF samples was determined by using a bomb calorimeter (Parr Calorimetric Thermometer-6772, Parr Instruments Company, USA).Briefly, energy of the dried material was determined by placing a weighed sample in the ignition cup of the bomb. Oxygen was filled at 25 psi and the bomb immersed in 2 L of water. The sample was ignited and the temperature increases of the water recorded. Gross energy is calculated using a boric acid calibration standard and expressed as Kcal g-1 (KJ K g-1) of dry matter

3.15 ANALYSIS OF FATTY ACID PROFILE

Extraction of lipid from the AMF sample was done by Folch method (Folch *et al.*,1957) with slight modifications. The extracted lipid was undergone FAME preparation and analyzed in GC-MS for fatty acid profile.

3.15.1 Extraction of Lipids

5 gram of sample was finely ground with a mixture of 20 ml chloroform and 10 ml methanol using homogenizer. The mixed solution is then filters through whattman No.1 filter paper into a separating funnel. Then 0.85 % NaCl is added to this and mixed properly and kept for separation into two layers. The lower chloroform layer with lipid is collected in a pre-weighted flat bottom flask and the solvent was evaporated using a rotary evaporator. Weight of the flask with lipid was taken and weight of lipid was calculated from the difference in the weight of the flask with lipid and without lipid.

3.15.2 Preparation of fatty Acid Methyl Esters (FAME)

The AOAC (1995) method was followed to esterify the lipid extract FAME was prepared from the lipids extracted from samples by heating with the methanoic NaOH first and then with BF3 Methanol for esterification.5 ml n-heptane was added to recover the methyl esters in an organic phase. Saturated NaCl solution was added to the mixture and the aqueous and organic layers were separated using a separating funnel,the upper n-heptane phase was pipet out and stored in 10ml glass vials in a refrigerator until further analysis.

3.15.3 Gas Chromatography-Mass Spectrometry

Fatty acids were separated using a Shimadzu Qp2010 Quadrupole Gas Chromatography Mass spectrometer (GC-MS) instrument equipped with a carbowax (30 m x 0.25 mm ID;0.25µm film thickness) capillary column (cromlab S.A). Helium was used as the carrier gas, Injector and detector temperatures were set at 250°C.injection was performed in split mode (1:15).The column temperature was programmed initially at 50°C for 2 minutes and then to increase at a rate of 10°C per min to a final temperature of 230°C. FAME esters were separated at constant pressure (23.1 kPa) and peaks were identified by comparing the mass spectra with the mass spectral database.

Table-6 Details of characteristic ions used in the study for identification of fatty acids

Class of fatty acid	Characteristic ion (m/z)
SAFA	74
MUFA	M-74
PUFA	91
n-6	150
n-3	108

3.16 CONFOCAL MICROSCOPY

The fluorescent images of bacterial floc were taken using a confocal microscope (Leica TCS 4PI) on a biweekly basis. Water samples were collected from one tank in each treatment and processed immediately for microscopy. Each sample was diluted with pre-filtered water of the same tank, stained with fluorescence 4'-6-Diamidino-2-phenylindole (DAPI), observed and photographed at magnification 10.

3.17 IDENTIFICATION OF PLANKTON IN AMF COMPOSITION

The plankton samples were collected in duplicate and concentrated to 50 ml by filtering the water using boiling silk cloth (no.25) from the respective experimental groups. The collected plankton samples were presented in 5 % formalin for further analyses (Phillpose.,1959;Pennak.,1978).Organisms were identified using keys and monographs given by Edmondson.,1992;Lund and Lund .,1995;Desikacharya., 1959,Graham *et al.*,2008,Prescott.,1969,APHA.,2005 and Adoni.,1975.The zooplankton was identified following Bradford –Grieve *et al.*(1999).The photographs of all the major planktons were taken using a camera attached to a Hund microscope(H600L,HundWetziar,Germany).

3.18 MICROBIOLOGICAL PARAMETERS

For a microbiological study, the samples were collected from the respective systems for every ten days during morning hours of the day between 8.00 to 10.00 a.m. The microbial floc suspensions were taken in the Imhoff cone and allowed to settle for 15-20 minutes. The settled AMF suspension was transferred aseptically into sterile Uricol (Hi-media) bottles and stored in a refrigerated condition.

3.18.1 Enumeration of total Heterotrophic Bacteria

Total Bacillus count was enumerated on nutrient agar medium from both systems. 10 ml of the floc water sample was transferred aseptically to the 90 ml physiological saline which gives 10 times dilution of the sample, i.e 10^{-1} dilution for further dilution,1 ml from 10^{-1} was mixed with 9 ml saline (10^{-2}) and so on to get the appropriate dilution for plating. Three dilutions were selected for plating in triplicate. 0.1 ml of each of the dilutions was inoculated on the surface of the respective pre-set agar plates. By using glass spreader the inoculums was uniformly speeded well on the surface of the agar medium. After 30 minutes, the plates were incubated at 37°C for 48 hrs. After 48 hours of incubation, the colonies were enumerated using a colony counter. The average colony counts of triplicate plates were taken for the total heterotrophic bacterial count and expressed as colony forming unit per ml (CFU ml^{-1}). Similarly vibrio count and fungi count from raceways and lined pond water samples was determined by the spread-plate technique Thio sulfate citrate bile salts sucrose (TCBS) agar for Vibrio count, Sabouraud dextrose agar for total fungal count were used respectively. A total of 10 isolates of bacteria from aerobic microbial floc in both systems were subjected to biochemical tests for finding the relative difference to categorize the bacterial groups, out of which five isolates were given for 16s r DNA amplification.

3.18.2 DNA extraction from AMF

Microbial suspension (1ml) was kept in Hi-Bead tubes (Hi-Media Laboratories Limited, Mumbai) and the lysozyme-SDS based phenol-chloroform method of extraction was followed with slight modifications 500 µl of lysis 1 solution (0.15 M NaCl, 0.1 M EDTA; pH 8; 15mg lysozyme/l) was added to the bead tube and mixed by horizontal vortexing for 2 minutes. It was then incubated at 37°C for 1 hour in a water bath. After that 500 µl of lysis 2 (0.1 M NaCl,0.5 MTris-HCL;pH8 :12 % SDS) solution was added to it and incubated again for 1-2 hours at 60°C in a water bath. The suspension was then centrifuged at 8000 g for 10 minutes in a Spinwin (Tarson, India) centrifuge and 1 ml of supernatant was recovered in separate vials. To this supernatant,1 ml of saturated phenol (pH 8) was added, mixed and centrifuged at 8000 g for 8 minutes. The aqueous layer was recovered and an equal volume of chloroform was added, mixed and centrifuged for 30 minutes at 4°C at 8000 g in a microprocessor –based high speed research refrigerated centrifuge (Eltek, India). The supernatant was decanted and pellets were washed with 500 µl of 70 % ethanol and centrifuged at 8000 g for 15 minutes using the centrifuge. Finally ,the vials were decanted again and kept in a dry bath (SLM-DB-120, Genei, India) at 60°C for some time to evaporate the remaining ethanol. The DNA was then re-suspended in 50 µl of nuclease-free water and stored at -20°C.

3.18.3 PCR Amplification using Universal Primers

The elute was checked for the presence of genomic DNA using 1.5 % agarose gel pre-stained with ethidium bromide (0.5 µg ml-l),1.0 µl of 6x loading dye was mixed with 10.0 µl of genomic DNA and loaded into the wells of the gel. Loading dye 1.0µl was mixed with 5.0 µl 1Kb DNA molecular weight marker and loaded into the wells. Electrophoresis was carried out in 1 x TAE buffer at 70 volts for about 60 minutes. The gel was visualized and photographed using Bio-Rad Universal Hood 11 Gel Imager (Bio –Rad Laboratories,USA).

Table- 7 Primer sequences used in the study

S.No	Primer Sequence (5'→3')
1.	27F5'-AGAGTTTGATCMTGGCTCAG-3'
2.	1492R5'-TACGGYTACCTTGTTACGACTT-3'
3.	518F 5'-CCAGCAGCCGCGGTAATACG-3',
4.	800R 5'-TACCAGGGTATCTAATCC-3')

Table- 8 Components of PCR reaction

S. No	Reagents	Volume(50 µl)
1.	Master mix	25
2.	Forward primer (100 picomoles)	0.4
3.	Reverse primer (100 picomoles)	0.4
4.	Template	3
5.	Nuclease-free water	21.2

3.18.4 Agarose Gel Electrophoresis for PCR Products

The PCR consisted of 30 cycles each of initial denaturation at 95°C for 5 minutes, denaturation at 95°C for 30 seconds, annealing at 61.3°C for 30 seconds and extension at 72°C for 1 minute,and a final extension for 15 minutes. The PCR products were examined by electrophoresis on 1 % agarose gel prestained with ethidium bromide and visualized in Bio-Rad Universal Hood 11 Gel Imager (Bio-Rad Laboratories,USA).

The PCR products were tested for the various biochemical tests such Gram staining, Observation of bacterial motility by hanging drop technique, 0/129 sensitivity (2,4 diamino 6,7 di-isopropyl pteridine), Lysine test, Ornithine test,

Urease test, Phenyl analine test, Nitrate reduction test, H2S test, Citrate test, Vokes Proskaeur (VP) test, Methyl red test, Indole test, Melanote test and Indole test as per APHA 2005.

3.18.5 Biochemical reaction tests

3.18.5.1 Motility

A darting zig – zag movement indicates polar flagellation and less vigorous and more vibratory movement indicates peritrichate flagellation by examining the bacteria under microscope

3.18.5.2 Gram Staining

The pararozaniline dye such as crystal violet treated with iodine mordant remains trapped in the cell wall and hence, the cells are not destained when treated with alcohol. Gram positive bacteria stain violet/blue while the gram negative.

3.18.5.3.ONPG Test

Cultures were grown on basal nutrient agar containing 1% (w/v) lactose to induce β–d–galactosidase, which was then detected using the toluene modification method of ONPG hydrolysis.

3.18.5.4 Lysine decarboxylase test

An inoculum from a pure culture is transferred aseptically to a sterile tube of lysine decarboxylase broth. The inoculated tube is incubated at 35 -37 C for 24 hours and the preliminary results are determined. The microbe must first use the glucose present to cause the pH to drop. This is indicated by a change from purple to yellow.

3.18.5.5 Ornithine test

An inoculum from a pure culture is transferred aseptically to a sterile tube of ornithine decarboxylase broth. Once the medium has been acidified, the enzyme ornithine decarboxylase is activated. The culture is incubated *an additional 24 hours* at 35 -37 C to allow the microbe to now use the ornithine. Change of colour from yellow to purple indicates a positive test for ornithine decarboxylase.

3.18.5.6 Urease Test

Urea slants are prepared and inoculated with culture to be tested. Observe the tubes for colour change after incubation for 24 h. Positive tubes turn to pink colour and negative tubes remain yellow.

3.18.5.7 Phenylalanine deaminase test

Phenylalanine deaminase medium tests the ability of an organism to produce the enzyme deaminase. After incubation, 10% ferric chloride is added to the media; if phenylpyruvic acid was produced, it will react with the ferric chloride and turn dark green.If the medium remains a straw color, the organism is negative for phenylalanine deaminase production.

3.18.5.8 Nitrate Reduction Test

The bacteria in a medium containing nitrate should be added with 0.1 ml of the reagent such as Solution A-8.0g sulphanilic acid in 1l of 5 N acetic acid and Solution B-5.0g naphthyl amine in 1l of 5 N acetic acid. Positive cultures show a development of red colour within few minutes.

3.18.5.9.H2S test

An inoculum from a pure culture is transferred aseptically to a sterile triple sugar iron agar (TSIA) slant. The inoculated tube is incubated at 35 -37 C for 24 hours and the results are determined. In tubes of TSIA containing bacteria producing hydrogen sulfide, the agar turns black from the FeS.

3.18.5.10.Citrate Utilization Test

Inoculate the culture to the medium and observe the colour change as Prussian blue colour. Make up the medium in distilled water. Melt and distribute into 3 ml quantities and autoclave at 121 C

3.18.5.11 Voges Proskauer (VP) test

Organism to be tested is inoculated in glucose phosphate peptone broth. Incubate at 37 C for 48-72 hours. To about 1-2.5 ml of culture add 0.6 ml of naphthol and 0.2 ml of 40% KOH. To provide more guanidine groups add a pinch of creatinine. Allow it to stand for 30 minutes to 2 hours. Positive reactions give the development of port wine colour. Usually MR positive organisms are VP negative and vice versa.

3.18.5.12 Methyl red Test

Organism to be tested is inoculated in glucose phosphate broth (GPB) and incubated at 37 C for 48 – 72 hours. After the incubation, add a few drops of the alcoholic solution of methyl red, and read the result immediately. Positive result gives red colour where as negative result is indicated by yellow colour

3.18.5.13 Indole test, Kovac's method

Incubate the organism in tryptone broth for 48 hours. Add Kovac's reagent and observe for colour. This reaction gives pink colour and the reagent will form a ring at the surface of the culture since P-DMAB solution is prepared in amyl alcohol.

3.18.5.14 Malonate Test

An inoculum from a pure culture is transferred aseptically to a sterile tube of malonate broth. The inoculated tube is incubated at 35 -37 C for 24 hours and the results are determined. Abundant growth and a change from green to blue in the medium indicates a positive test for growth using malonate.

3.18.6 Sequences analyses

Sequences were analyzed for similarity with sequences deposited in public databases using the Basic Local Alignment Search Tool (BLAST) (McGinnis & Madden 2004)

at the National Center for Biotechnology Information database (http://www.ncbi.nlm.nih.gov/BLAST). Alignment and further phylogenetic analysis were performed by means of Molecular Evolutionary Genetics Analysis (MEGA) software version 4.0 (Tamura, Dudley, Nei & Kumar 2007). The resulting alignment were manually checked and corrected when necessary and unambiguously aligned nucleotide position were used for construction of a 16S rDNA sequence based phylogenetic tree, using the neighbor-joining (NJ) method (Saitou & Nei 1987) because of its known accuracy. The tree is drawn to scale, with branch lengths in the same units as those of the evolutionary distances used to infer the phylogenetic tree. The evolutionary distances were computed using the Maximum Composite Likelihood method (Tamura, Nei & Kumar 2004) and are in the units of the number of base substitutions per site.

The closely related sequences were stored at -80°C and directly sent for sequencing to Chennai, Genei, India. The partial bacterial 16SrDNA gene sequences were subjected to BLAST (www.ncbi.nlm.nih.gov/BLAST/) search on the Genbank nucleotide database NCBI (National Centre for Biotechnology Information) to identify the sequences with higher similarity.

3.19 TECHNO - ECONOMIC FEASIBILITY OF AMF DRIVEN NURSERY RACEWAYS AND LINED POND CULTURE

This study had documented the techno –economic feasibility of the two systems by adopting Felix.S.,2014.The fixed cost and variable cost were calculated according to the input prices in Tamilnadu, where the justifications behind the economics of the production quantity are transparent to attach credibility to the culture systems and establish confidence in the outcome of the study.

4 CHAPTER

RESEARCH DATA FINDINGS

4.1 PHYSICO-CHEMICAL PARAMETERS OF WATER

4.1.1 Temperature

The water temperature of the different experimental groups at 06:00 hrs, 12:00 hrs,18:00 hrs ranged from 26.0°C to 29.0 °C,29.0°C to 32.0 °C and 28.0°C to 29.0 °C . in indoor raceways and in outdoor lined ponds it ranged from 25.0°C to 29.0 °C , 31.0°C to 36.0 °C, 30.0°C to 33.0 °C during the experimental period. There was no significant difference($P>0.05$) in temperature among between the system (Table-9).

4.1.2 pH

The fluctuation of pH was observed at 06:00,12:00 and 18:00 hrs ranged from between 7.2 to 7.6 ,7.2 to 7.8 and 7.2 to 7.8 in indoor raceways .In lined ponds the values were ranged from 7.2 to 7.9,7.0 to 7.9 and 7.3 to 7.9 during the experiment. There was no significant variation ($P>0.05$) in the mean pH between the experimental groups (Table-9).

4.1.3 Total alkalinity

The observations presented in table-9 shows that there was no significant difference ($P>0.05$)in the total alkalinity between the systems. The mean total alkalinity ranged from 95 to 178 mg $CaCO_3$ L^{-1} in raceways and from 110 to 172 mg $CaCO_3$ L^{-1} in lined ponds.

4.1.4 Calcium hardness

The calcium hardness in the water sample did not vary significantly ($P>0.05$) between raceways and lined ponds. The values were found in the range of 117-280 mg $CaCO_3$ L^{-1} in raceways and 102-155 mg $CaCO_3$ L^{-1} in lined ponds (Table.9).

4.1.5 Magnesium hardness

The Magnesium hardness in the water sample did not vary significantly ($P>0.05$) between raceways and lined ponds. The values were found in the range of 106-315 mg $CaCO_3$ L^{-1} in raceways and 102-155 mg $CaCO_3$ L^{-1} in lined ponds. (Table.9).

4.1.6 Total hardness

The total hardness was found to be 1120 to 2220 mg $CaCO_3$ L^{-1} in raceways and 840-1650 in lined ponds during the experimental period. There was no significant variation ($P>0.05$) of the values in raceways and lined ponds (Table 9).

4.1.7 Dissolved Oxygen

The concentration of dissolved oxygen in the two systems were recorded within the range of 3.8 to 5.6 ,4.0 to 6.8,3.8 to 8.4,5.5 to 8.4,5.0 to 8.2 and 4.0 to 8.2 mg L^{-1} in Indoor raceways at 01:00,04:00,09:00,13:00,16:00 and 19:00 hrs respectively. In Outdoor lined ponds 3.5 to 5.9,3.8 to 5.5,4.0 to 8.0,5.2 to 7.2,4.8 to 6.8 and 4.5 to 6.8 mg L^{-1} during the experimental period. Dissolved oxygen showed no significant variation ($P>0.05$) between the raceways and lined ponds (Table.9) .

4.1.8 Carbon dioxide

The Carbon dioxide concentration level between the systems were not significantly different ($P>0.05$) and in the indoor raceways (Table.9) in the range of 0.5-5.12 mg L^{-1} whereas outdoor ponds was in the range of 0.9-3.10 mg L^{-1}

4.1.9 Ammonia –Nitrogen (NH3-N)

The concentration of ammonia nitrogen at 04:00,14:00 and 23:00 hrs ranged from 0.03 to 0.48,0.04 to 0.68 and 0.02 to 0.43 mg L^{-1} in raceways and from 0.01 to 0.9,0.01 to 1.2 and 0.01-1.3 mg L^{-1} in lined ponds .The concentration of TAN vary significantly between the systems i.e there was a significant variation ($P<0.05$) in raceways and lined ponds. (Table 9)

4.1.10 Nitrite-Nitrogen (NO2-N)

The nitrite-nitrogen concentration fluctuated from 0.01 to 0.15 mg L^{-1} in raceways and 0.01 to 1.71 mg L^{-1} in lined ponds during the experimental period, the concentration of nitrite-N among the treatments did not vary significantly ($P>0.05$) (Table 9).

4.1.11 Nitrate-Nitrogen (NO3-N)

The Nitrate-nitrogen concentration was recorded within the range of 3 to 45 mg L^{-1} in raceways and from 12 to 257 mg L^{-1} during the experimental period.The concentration of nitrate-N between the systems varied significantly ($P<0.05$) (Table 9).

4.1.12 Total Phosphate

The total phosphate concentration in the raceways 1.65-12.40 mg L^{-1} and in lined ponds 1.59-32.40 mg L^{-1} .There was a significant difference ($P<0.05$) between the raceways and lined ponds(Table.9) .

4.1.13 Sodium

There was no significant difference between raceways and lined ponds ($P>0.05$) and the range of sodium concentration in the raceways was 147-165 mg L^{-1} and the lined ponds (Table.9) was 123-150 mg L^{-1}

4.1.14 Potassium

The concentration of potassium was in the range of 77- 228 mg L^{-1} in raceways and 100- 195 mg L^{-1} in outdoor lined ponds. There was no significant difference ($P>0.05$) between raceways and lined ponds (Table.9).

4.1.15 Salinity

The salinity maintained in the range of 6-7 ppt throughout the experiment in both the treatments. Hence there was no significant difference ($P>0.05$) between raceways and lined ponds throughout the experiment (Table.9).

4.1.16 Total Suspended solids

The TSS fluctuation ranged from 124 to 237 mg L^{-1} in the raceway and from 85 to 139 mg L^{-1}.The TSS in lined ponds and raceways vary significantly ($P<0.05$) between raceways and lined ponds(Table.9).

4.1.17 Volatile suspended solids

The VSS concentration fluctuated between 94 to 131 mg L^{-1} in raceway and from 47 to 87 mg L^{-1} in lined pond during the study period. The VSS in lined ponds and raceways have a significant difference ($P<0.05$) between them(Table 9).

4.1.18 Total Dissolved Solids

The Total Dissolved Concentration in the raceways was found to be 6.40-15.60 mg L^{-1} and in lined ponds it was 2.25-8.50 mg L^{-1}.The TDS in raceways and lined ponds vary significantly ($P<0.05$) presented in the table.9.

4.1.19 Turbidity

The turbidity level in the raceways 26.6 - 79.2 NTU and 28.4-41.3 NTU in lined ponds (table.9).There was a significant difference between raceways and lined ponds($P<0.05$)

4.1.20 Biochemical oxygen demand

BOD in raceways was found to be 33.21-145.20 mg L^{-1} and in lined ponds it was 51.44 -121.40 mg L^{-1}. There was a significant difference ($P<0.05$) between raceways and lined ponds (Table.9).

4.1.21 Floc volume

Floc volume of Aerobic Microbial Floc in raceways (8.4-12.3 ml L^{-1})was significantly higher than lined ponds(2.6-8.00 ml L^{-1}).They vary significantly ($P<0.05$) between raceways and lined ponds (Table-9)

Table 9.Water quality parameters of nursery raceways and lined ponds during the nursery trial of 50 days

Parameters	Time (hrs)	Raceways(R)	Lined Ponds(P)
Temperature(˚C)	06:00	27.25±0.64 (24-29)	26.71±1.01 (23-29)
	12:00	30.32±0.61 (29-32)	33.14±1.43 (31-36)
	18:00	28.78±0.41 (28-29)	31.07±0.76 (30-33)
pH	06:00	7.5±0.14 (7.2-7.6)	7.5±0.19 (7.2-7.9)
	12:00	7.4±0.14 (7.2-7.8)	7.5±0.21 (7.0-7.9)
	18:00	7.4±0.13 (7.2-7.8)	7.6±0.19 (7.3-7.9)
Alkalinity (mg $CaCO_3$ L^{-1})	11.00	122.14±5.69 (98-178)	149.25±5.99 (110-172)
Calcium Hardness (mg $CaCO_3$ L^{-1})	11.00	156.60±6.79 (117-280)	134.71±7.98 (102-155)
Magnesium Hardness (mg CaCO3 L^{-1})	11.00	167.96±9.74 (106-315)	144.53±7.88 (100-195)
Total Hardness (mg CaCO3 L^{-1})	11.00	1906±229.22 (1120-2220)	1178±198.68 (840-1650)
Dissolved Oxygen (mg L^{-1})	01:00	5.52±0.03 (3.8-5.6)	4.83±0.01 (3.5-5.9)
	04:00	5.0±0.04 (4.0-6.8)	4.2±0.02 (3.8-5.5)

	09:00	6.4±0.01 (3.8-8.4)	5.57±0.05 (4.0-8.0)
	13:00	6.2±0.04 (5.5-8.4)	6.01±0.03 (5.2-7.2)
	16:00	5.9±0.01 (5.0-8.2)	6.1±0.03 (4.8-6.8)
	19:00	5.9±0.08 (4.0-8.2)	5.01±0.04 (4.5-6.8)
CO_2 (mg L^{-1})	11.00	15.53±0.25[a] (2.3-18.4)	13.16±0.2[a] (4.4-15.2)
NH_4-N(mg L^{-1})	04:00	1.2±0.04[a] (0.3-1.5)	1.4±0.08[b] (0.1-1.8)
	14:00	0.9±0.06[a] (0.4-0.9)	0.7±0.06[b] (0.1-1.5)
	23:00	0.46±0.03[a] (0.2-0.7)	0.6±0.03[b] (0.1-1.3)
Nitrite-NO_2-N (mg L^{-1})	11.00	0.55±0.07 (0.01-0.15)	0.68±0.09 (0.01-1.71)
Nitrate-NO_3-N (mg L^{-1})	11.00	24.78±2.59[a] (3-45)	68.28±5.17[b] (12-97)
Phosphate (mg L^{-1})	11.00	5.61±0.71[a] (1.65-12.40)	8.14±1.27[b] (1.59-18.20)
Sodium (mg L^{-1})	11.00	160.4±4.50 (147-165)	142.4±6.80 (123-150)
Potassium (mg L-^{1})	11.00	147.96±15.74 (77-158)	144.53±12.88 (100-195)
Salinity (‰)	11.00	6.9 ± 0.5 (6.5–7.7)	6.8 ± 0.4 (6.4-7.2)
TSS (mg L^{-1})	11.00	155.87±20.26[a] (124-237)	116.75±14.63[b] (85-139)
TDS(mg L^{-1})	11.00	175.01±1.50[a] (89-184)	102±0.55[b] (45-115)
VSS(mg L^{-1})	11.00	107.62±8.92[a] (94-131)	62.75±6.5[b] (47-87)

Turbidity (NTU)	11.00	44.59±7.50[a] (26.6-79.2)	35.23±1.58[a] (28.4-41.3)
BOD(mg L^{-1})	11.00	92.28±11.72[a] (33.21-145.2)	81.85±9.91[b] (51.44-121.40)
Floc Volume (ml L^{-1})	11.00	12.3±1.12[a] (8.4-12.30)	6.08±1.03[b] (2.6-8.00)

TAN : Total Ammonia-N ; **TSS**: Total Suspended Solids; **VSS**: Volatile Suspended Solids ; **BOD** :Biochemical Oxygen Demand.

Values are means (±SE, n= 20) in two replications at regular interval for indoor raceways and outdoor lined ponds. Mean values in the same row with different superscript differ significantly (P<0.05)

4.2 WATER EXCHANGE RATE

The AMF system in indoor raceways saved about nearly 3 times of the water needed to maintain nontoxic conditions during vannamei production in outdoor culture. Additional water to maintain AMF water quality in indoor raceways was 108m^3,while water to maintain AMF culture was 352 m^3 in outdoor ponds. The details of the total water consumption was presented in the table 10.

Table-10 Total Water Consumption throughout the experimental period of 50 days in nursery raceways and lined ponds

Specifications	Raceways	Lined ponds
Total volume	50	100
Initial fill (m^3)	40	80
Water replacement volume (m^3)	4 (@ 10%)	16 (@ 20%)
17 top offs (m^3)	68	272
Total water consumption(m^3)	108	352

4.3.GROWTH PARAMETERS

4.3.1 Average Body Weight (ABW)

The body weight of the experimental group was recorded for every ten days once as shown in Fig.1 .There was a significant difference (P<0.05) in the average body weight between the systems. The initial weight (g) was 0.024±0.002g in raceways and 0.027±0.001g in lined ponds. The final body weight (g) varied in the range of

0.907+0.057 in raceways and 1.198+0.086 in lined ponds (Table.11). The lined pond culture showed more weight gain compared to raceway culture.

4.3.2 Specific Growth Rate (SGR)

The value of SGR did not vary significantly (P>0.05) between the systems (Fig.3). The higher SGR value was found in the raceways (7.22±0.04) and lowest SGR (7.03±0.13) was found in the lined pond (Table.12)

4.3.3 Feed Conversion Ratio(FCR)

The feed conversion ratio value significantly (P>0.05) did not vary between the system (Table.12).The higher FCR was recorded in raceway treatment (1.17±0.06) and the lower FCR was found in lined ponds (0.89±0.03) (Fig.4).

4.3.4 Protein Efficiency Ratio (PER)

The mean PER value was significantly different (P<0.05) between the system(Fig.5) .The significantly higher PER value was recorded in lined ponds (3.87±0.09) and significantly lowest PER was recorded in raceway (2.94±0.12) (Table.12)

4.3.5 Yield

The yield of the cultured shrimp juveniles vary significantly (P<0.05) between raceways and lined ponds (Fig.6).There was a significant difference between raceways (4.92±0.26) and lined ponds (2.36±0.02) (Table.12).

4.3.6 Survival Rate

The survival rates of the animals between the experimental groups are given in the table 12 and fig.7.There was no significant difference (P>0.05) in survival rate between the raceways and lined ponds. The mean survival rate was found to be higher in raceways (91.2±1.6%) than the lined ponds (86.1±1.4 %).

Table-11 Average body weight (mean ± SE, n=2) of vannamei in nursery raceways and lined ponds

Days	Raceways	Lined Ponds
0	0.024^{a}±0.002	0.027^{b}±0.001
10	0.038^{a}±0.003	0.054^{b}±0.002
20	0.069^{a}±0.005	0.097^{b}±0.006
30	0.129^{a}±0.006	0.156^{b}±0.004
40	0.534^{a}±0.034	0.517^{b}±0.028
50	0.907^{a}±0.057	1.198^{b}+0.086

The mean values (g) in a column bearing different superscript differs significantly (P<0.05)

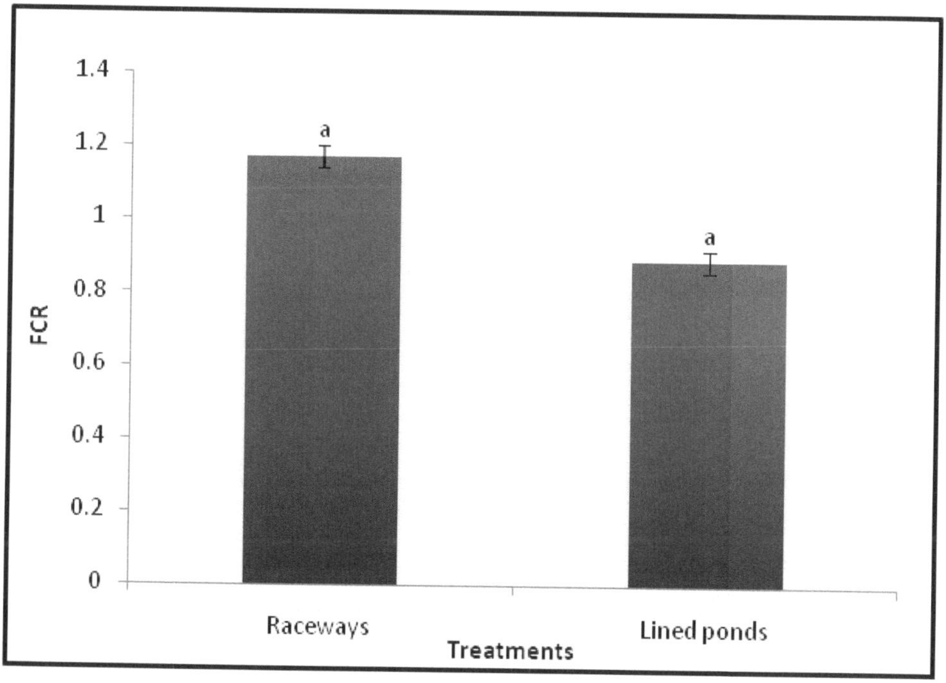

Fig 5.PER values (mean ±SE) in raceways and lined ponds.

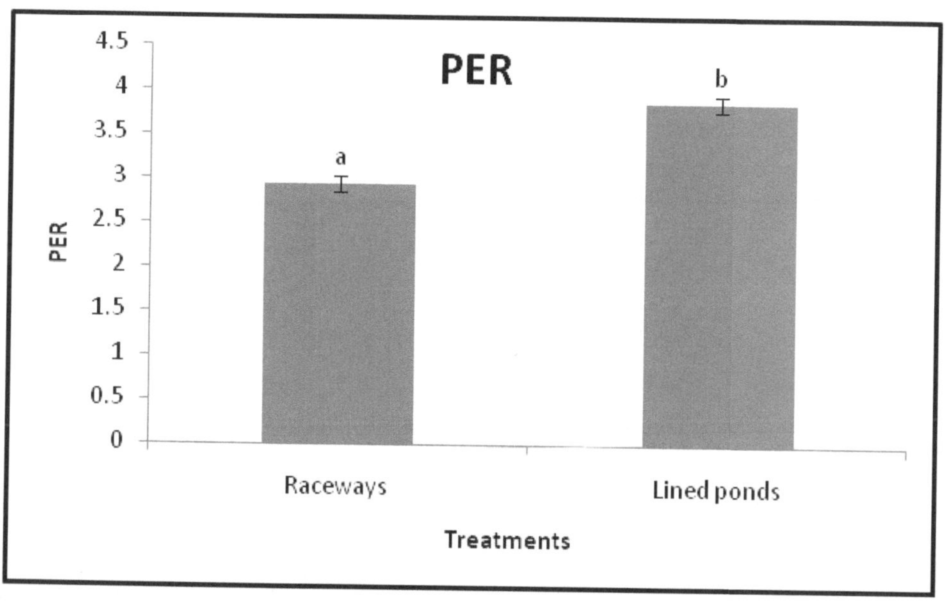

Fig.6.Yield values (mean ±SE) in raceways and lined ponds

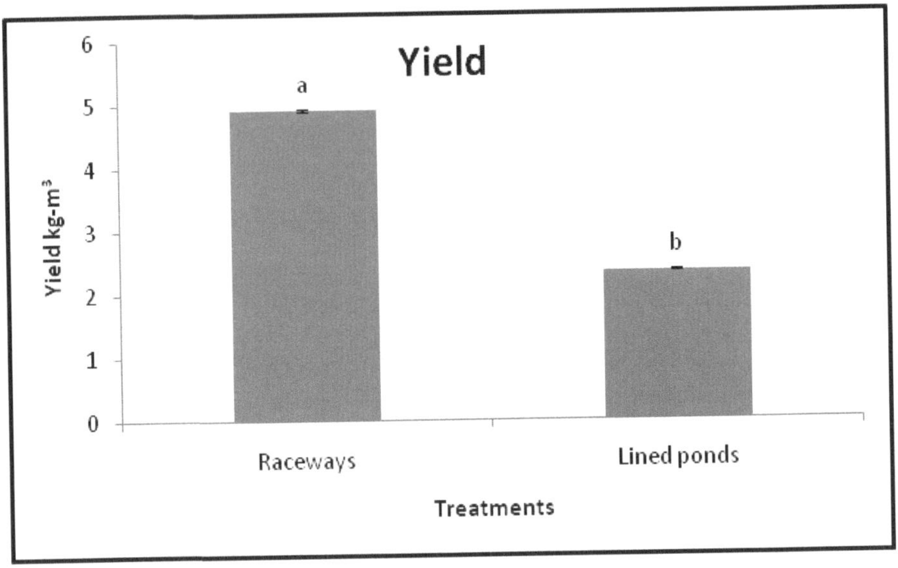

Fig.7. Survival percentage of (mean ±SE) in indoor raceways and outdoor lined ponds

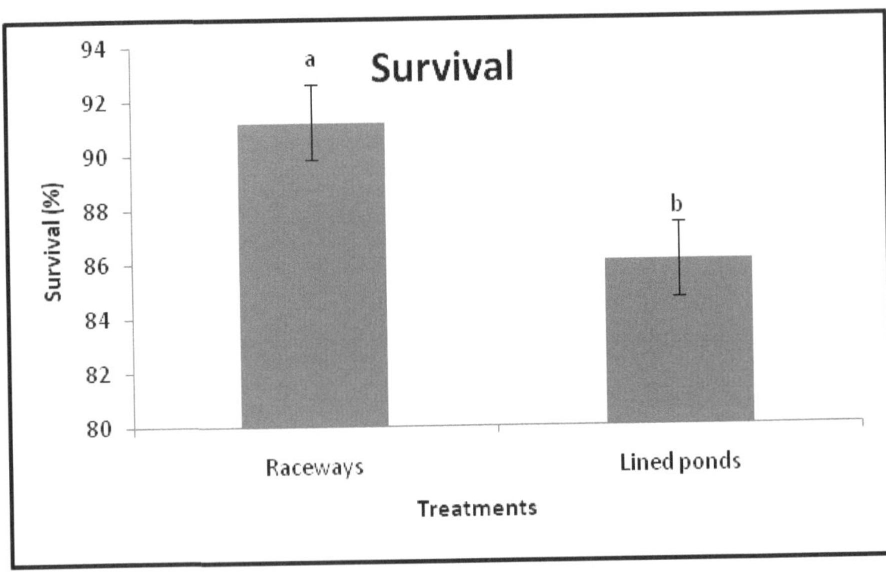

4.4 FEED ECONOMICS OF THE EXPERIMENTAL GROUPS

The costs for the production of 1 kg shrimp live weight produced are presented below:-

- (1.2 kg feed kg^{-1} fish live weight produced × ₹ 65 kg -1 feed) + (1.2 kg MF-CEED kg^{-1} fish live weight produced × ₹ 10 kg^{-1} MF-CEED) = ₹ 88 kg^{-1} fish live weight produced in Indoor raceways

- (0.9 kg feed kg^{-1} fish live weight produced × ₹ 65 kg^{-1} feed) + (1.2 kg MF-CEED kg^{-1} fish live weight produced × ₹ 10 kg^{-1} MF-CEED) = ₹ 69 kg^{-1} fish live weight produced in outdoor lined ponds

The gain appears to be in the order of 22% in terms of feed costs kg^{-1} shrimp live weight produced in outdoor ponds, compared to indoor raceway culture at higher stocking density .

4.5 NUTRITIONAL QUALITY OF AMF

4.5.1 Proximate composition of AMF

There were significant differences in Crude protein(29.82±0.60 vs 37.22±1.80) and Total ash (28.95±3.01 vs 45.5±0.5) between raceways and lined ponds. They were significantly higher in outdoor lined ponds. The crude fat(6.5±0.5 vs 4.4±0.1) was significantly higher in raceways than lined ponds. The crude fibre (4.71±0.62 vs 4.6±0.2) remains the same in both the culture systems. The Ether Extract, carbohydrate and Gross energy were not significantly different between raceways and lined ponds. However, the Ether Extract (1.215±0.02 vs 1.85±0.02) was relatively higher in lined ponds. The carbohydrate (23.12±1.31 vs 25.01±0.7) was comparatively higher in lined ponds than raceways. Similarly Gross energy (287.7±2.2 vs 229.5±1.5) content of Aerobic Microbial Floc in lined ponds was higher.The results are presented in the Table.13.

Table-13 The proximate composition of AMF collected from raceways and lined ponds (%mean± SE ,n=2) in 50 days trial.

Nutrients	Raceways	Lined Ponds
Crude protein	29.82±0.60^{a}	37.22±1.80^{b}
Crude lipid	6.5±0.5^{a}	4.4±0.1^{b}
Crude Fibre	4.71±0.32^{a}	4.6±0.2^{a}
Total Ash	45.5±0.5^{a}	28.95±3.01^{b}
Ether Extract	1.215±0.02^{a}	1.85±0.02^{a}
Total Carbohydrate	23.12±1.31^{a}	25.01±0.7^{a}
Gross Energy KJ g^{-1}	229.5±5.5^{a}	287.7±6.2^{a}

Mean values in the same row with different superscript differ significantly ($P<0.05$)

4.5.2 Fatty acid profile of Aerobic Microbial Floc

The relative contents of C18:2n6 (linoleic acid), C20:5n3 (eicosapentaenoic acid), C22:6n3 (docosahexaenoic acid), n-3 PUFA, and n-6 PUFA had no significant differences between the systems. Only C18:2 (Oleic acid) was significantly higher in indoor raceways than outdoor lined ponds. The C20:0 (arachidic acid) ,C16:1

(palmitoleic acid) and C18:2 n-3 (linolenic acid) were significantly higher in outdoor lined ponds than indoor raceways. The results are presented in the Table-14.

Table-14 The Fatty acid profile of AMF collected from from raceways and lined ponds (% mean± SE, n=2) in 50 days nursery trial.

Fatty acids (%)	Raceways	Lined Ponds
C14:0 –Myristic Acid	2.92±0.25[a]	3.68±0.04[a]
C16:0-Palmitic acid	29.69±1.18[a]	24.995±0.57[a]
C18:0-Stearic Acid	10.84±0.1[a]	5.155±0.17[a]
C18:2 –Oleic Acid	31.91±0.12[a]	21.165±0.98[b]
C18:2 n-6 –Linoleic Acid	13.235±1.1[a]	14.325±1.29[a]
C18:3 n-3–Linolenic Acid	1.5±0.1[a]	13.05±0.35[b]
C20:0 –Arachidic Acid	0.375±0.015[a]	1.81±0.19[b]
C22:0 –Behenic Acid	1.24±0.12[a]	1.445±0.10[a]
C20:5 n-3(EPA)	0.665±0.05[a]	0.78±0.05[a]
C22:6 n-3(DHA)	0.79±0.05[a]	1.035±0.08[a]
C16:1-Palmitoleic Acid	6.995±0.60[a]	9.465±0.51[b]
Others	0.84±0.02[a]	0.6±0.01[a]

Mean values in the same row with different superscript differ significantly ($P<0.05$)

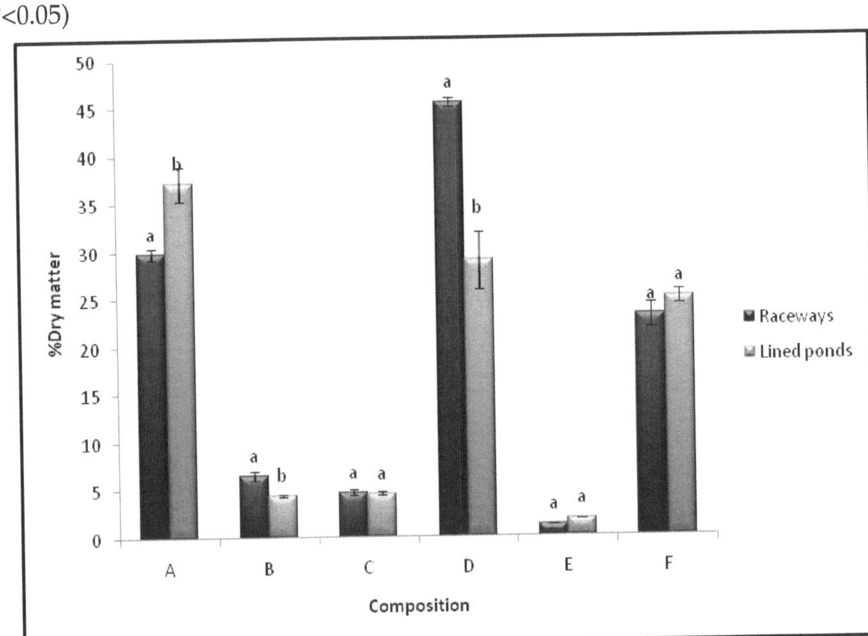

Fig.8 :Proximate composition of AMF in indoor raceways and outdoor lined ponds

A. Crude protein ; **B**. Crude Fat ; **C**. Crude Fibre; **D**. TotalAsh; **E**. Ether Extract; **F**. Total Carbohydrate

Mean values in the same row with different superscript differ significantly (P<0.05)

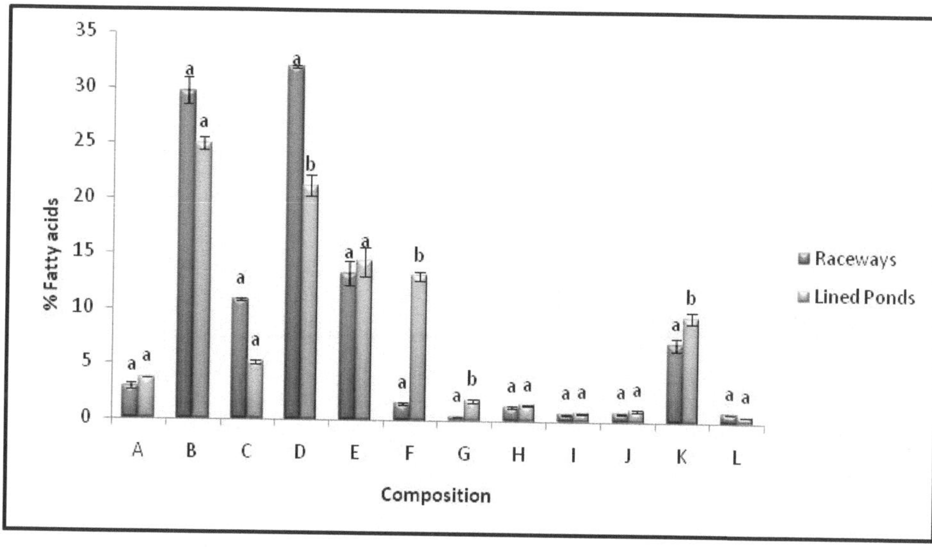

Fig.9 Fatty acid profile of AMF collected from raceways and lined ponds

Mean values in the same row with different superscript differ significantly (P<0.05)

4.8 CONFOCAL MICROSCOPY

A 3-D picture of individual bacterial flocs using a confocal microscope is shown in Fig.10.The blue channel dominating the picture indicates the clusters of bacterial nuclei. There are trace amounts of mucopolysaccharides, typically secreted by bacteria (green channel).

4.9 PLANKTON IDENTIFICATION AND COMPOSITION OF AMF

The structure of AMF was observed under a microscope with 10 x objective .In all experimental groups, anomalous flocculation with bacteria and zooplankton especially nematodes and rotifers were seen (Fig.11).The taxonomic composition of different floc-associated planktonic organisms are given in Fig10.In AMF, the planktons such as *Chlorella* sp., *Oscilatoria* sp., *Stephanodiscus* sp., *Coscinodiscus* sp., *Navicula* sp. (Pennate diatoms), *Scenedesmus,Vorticella, Gloenkinia, Chlorogonium,* dinoflagellates, *Amphiprora* sp., *Nitzschia* sp. (Pennate diatoms), *Chaetoceros* sp., *Cyclotella* sp., *Triceratium* sp., *Cymbella* sp., *Stentor* sp., *Paramecium* sp., *Cyclidium* sp., *Peranema* sp., *Petalomonas* sp., *Rotifer*, Nematode or round worm, *Chaetonotus* sp. *Cyclops* sp tintinids,cilites,copepods cyanobacteria and nematodes.Among the groups of organisms,nematodes were the most dominant in both systems. These microorganisms were found to be grazing on the flocs, when fresh samples were observed under microscope. However, predominantly the ciliates, protozoans and rotifer population was grazing within the AMF particles collected from indoor raceways compared to outdoor ponds. Some of the identified planktons are presented in the Fig.11.

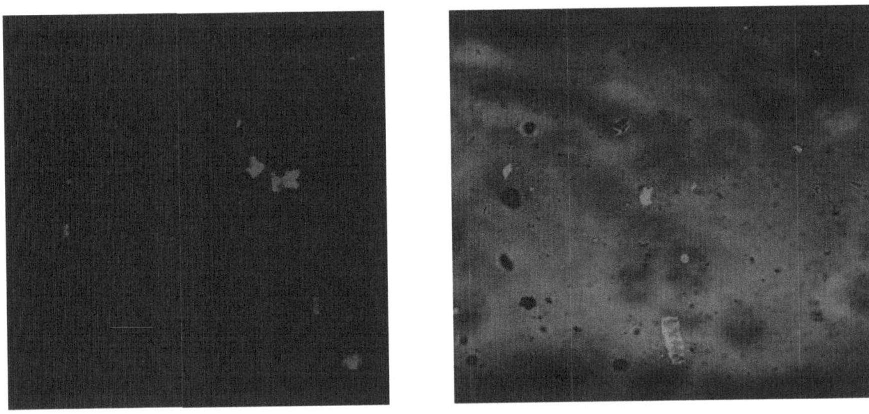

Fig. 10. Confocal microscopic images of AMF collected from raceways and lined ponds.

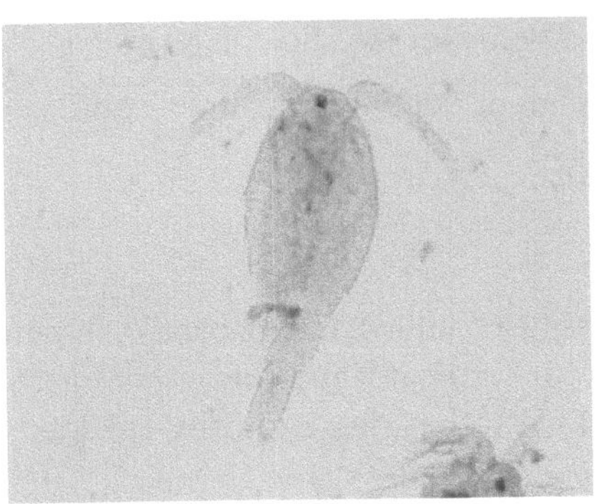

Copepod

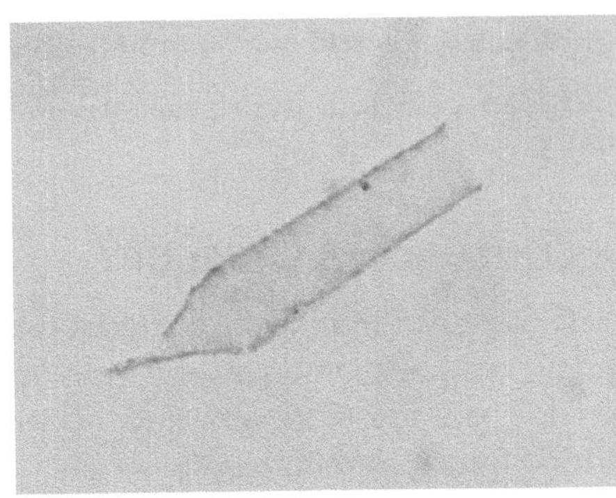

Tintinnids

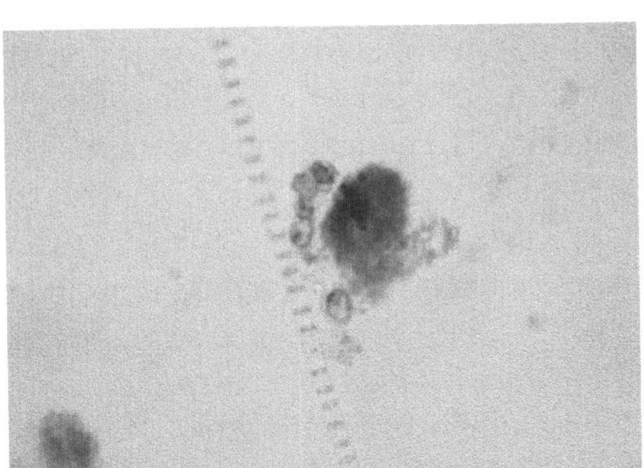

Skeletonema

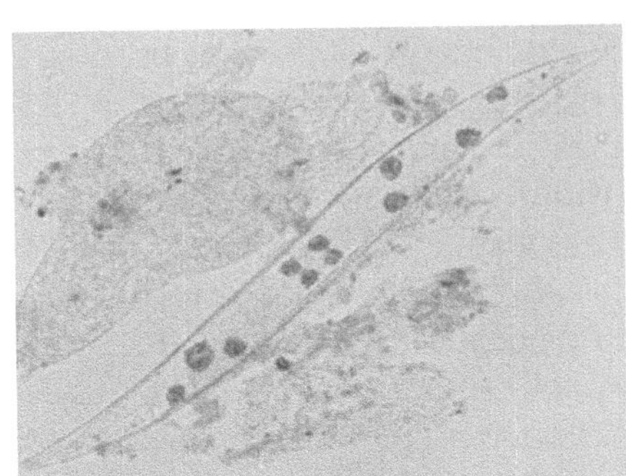

Pleurosigma

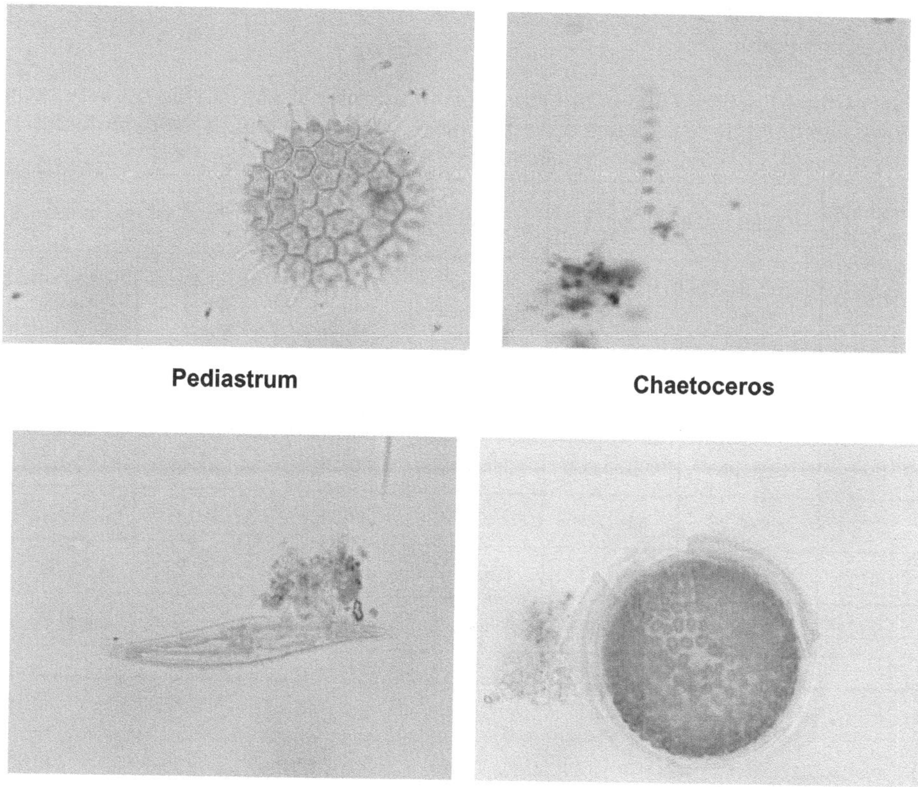

Pediastrum Chaetoceros

Navicula Coscinodiscus

Fig.11. Common Zooplankton and phytoplankton observed in the AMF driven nursery raceways and lined ponds

4.10 MICROBIOLOGICAL PARAMETERS

4.10.1 Total Bacillus Count

The total bacillus count population during the experimental period, ranged from 0.03±0.002x10^5 to 210±0.27x10^5 CFU ml-l in Raceways and 0.01 ±0.001x10^5 to 150±0.26x10^5 CFU ml^{-1} in lined ponds. They differ significantly ($P<0.05$) between the raceways and lined ponds (Table.15).

4.10.2 Total Fungal Count

The total fungal count was 1.5±0.14x10^3 to 21±0.11x10^3 CFU ml^{-1}in raceways and 0.3±0.5x10^3 to 11±0.36x10^3 CFU ml^{-1} in lined ponds and they differ significantly ($P<0.05$) (Table.15).

4.10.3 Total Vibrio Count

The Vibrio Count in the Raceways were found in the range of 1.1±0.03x10^3 to 5.21±0.25x10^3 CFU ml^{-1} and in the lined ponds 1±0.22x10^3 to 7.4±0.22x10^3 CFU ml^{-1}

respectively. There was a significant difference (P<0.05) between the raceways and lined ponds (Table.15).

Table-15 Total Bacillus,Fungal and Vibrio count of water in AMF driven nursery raceways and lined ponds. Values are mean (±SE ,n=2) at the end of the experiment.

Sampling Interval	Total Bacillus Count		Total Fungal Count		Total Vibrio Count	
Days	Indoor Raceways Mean (10^5CFU/ml)	Lined Ponds Mean (10^5CFU/ml)	Indoor Raceways Mean (10^3CFU/ml)	Lined Ponds Mean (10^3 CFU/ml)	Indoor Raceways Mean (10^3 CFU/ml)	Lined Ponds Mean (10^3 CFU/ml)
10	0.03±0.002 [a]	0.01±0.001 [b]	1.5±0.14 [a]	0.3±0.09 [b]	1.1±0.03 [a]	1±0.08 [b]
20	13±0.25 [a]	0.09±0.009 [b]	2±0.22 [a]	0.9±0.10 [b]	1.7±0.16 [a]	1.8±0.13 [b]
30	22.0±0.26 [a]	4±0.24 [b]	4±0.25 [a]	2±0.11 [b]	1.95±0.16 [a]	2.5±0.22 [b]
40	130±5.78 [a]	11±0.38 [b]	12±0.28 [a]	8±0.32 [b]	2.35±0.22 [a]	4.4±0.22 [b]
50	210±8.77 [a]	150±7.26 [b]	21±1.11 [a]	11±0.36 [b]	5.21±0.25 [a]	7.4±0.22 [b]

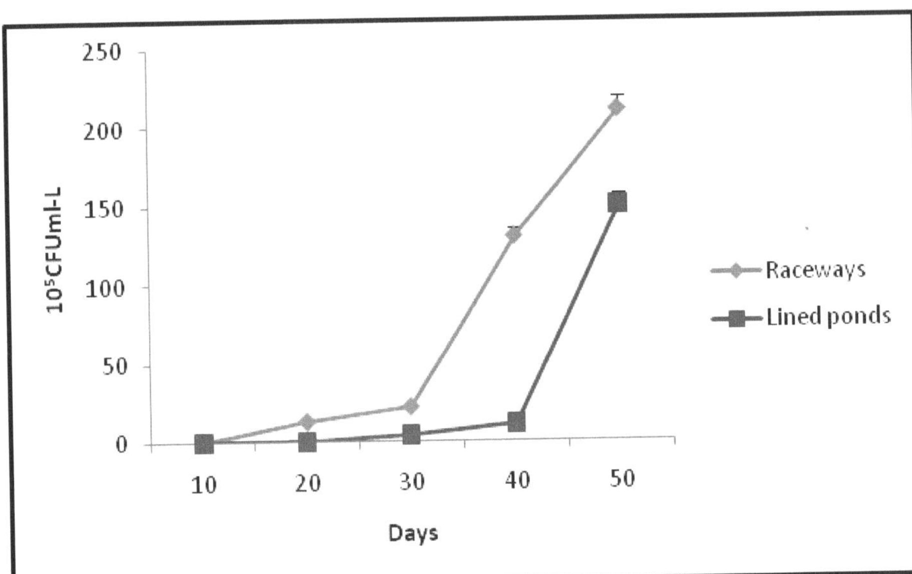

Fig-12Total Bacillus Count in raceways and lined ponds for 50 days

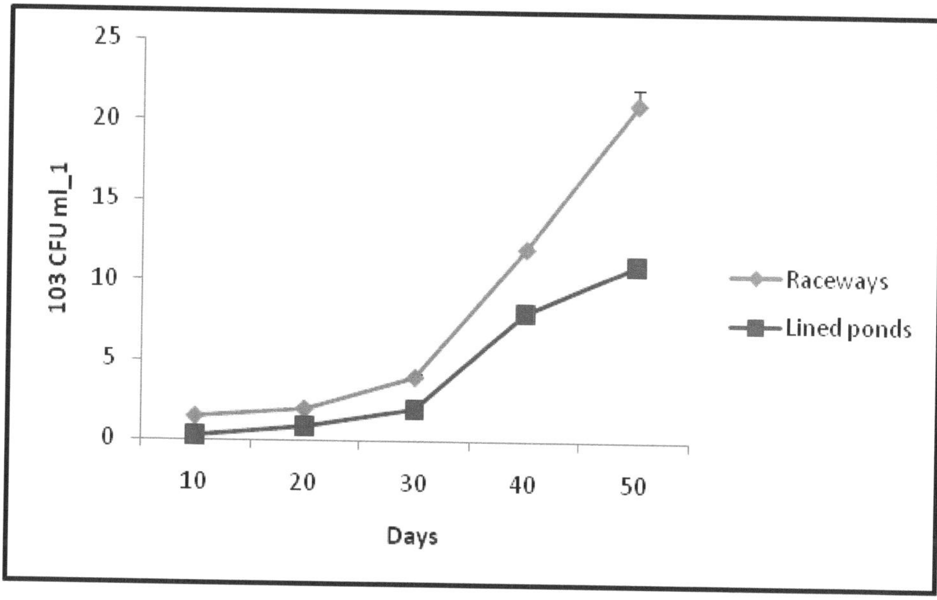

Fig.12 Total Fungal Count in raceways and lined ponds for 50 days

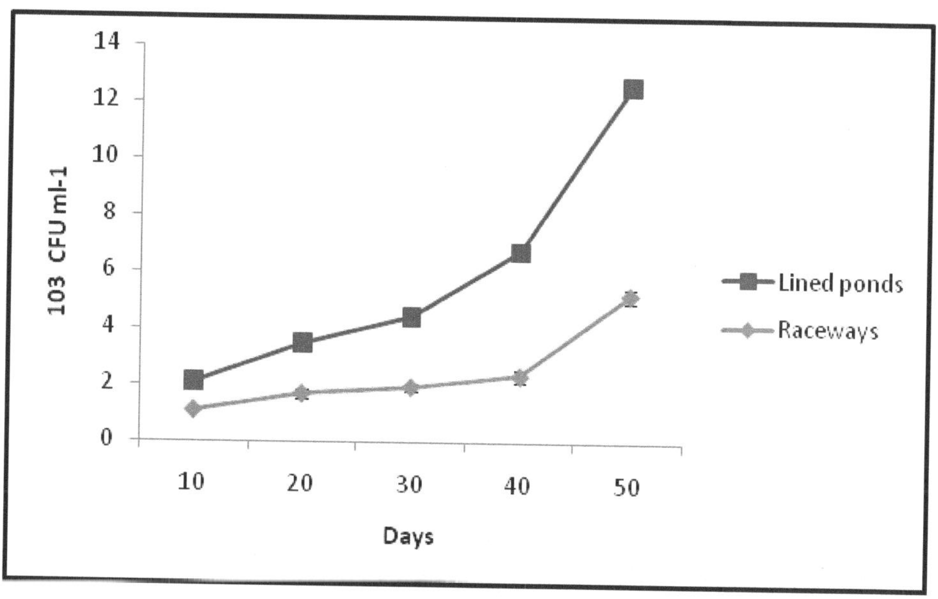

Fig.13 Total Vibrio Count in raceways and lined ponds for 50 days

4.11 BACTERIAL DIVERSITY

To investigate the bacterial diversity of Aerobic Microbial Flocs, the genomic DNA extracted from the flocs were subjected to 16S rDNA gene by Polymerase Chain

Reaction. Totally ten bacterial isolates have been selected from the culture plates (five from raceway culture and five from lined pond culture) and were tested for the several biochemical tests. The results of the biochemical test performance of the ten bacterial isolates are presented in the table 16.Among which R1, R2, R5, P2, P4 isolates were selected for PCR Amplification based on the biochemical test results. The gel images obtained after running agarose gel electrophoresis are given in figure.14.The PCR products were then purified and sequenced. Comparison of the 16 S rDNA gene sequence, derived from the Agarose gel bands with those of GenBank database revealed the identity of the organisms

Table.16 Observation of Biochemical test of bacterial isolates from AMF based nursery culture of *Litopenaeus vannamei.*

S.No.	Characterization	R1	R2	R3	R4	R5	P1	P2	P3	P4	P5
1	Motility	+	+	+	-	+	-	+	+	+	-
2	Gram staining	+	+	+	-	+	+	+	-	-	+
3	ONPG	-	-	-	-	-	-	-	-	-	-
4	Lysine	-	-	-	-	-	-	-	-	-	+
5	Ornithine	+	-	-	-	+	-	-	-	+	+
6	Urease	-	-	-	-	+	-	-	-	-	-
7	Phenylalanine	+	-	-	-	-	-	-	-	-	-
8	Nitrate reduction	+	+	+	+	+	-	+	+	+	+
9	H_2S	+	+	+	-	+	-	+	-	+	-
10	Citrate	-	-	-	+	-	-	-	+	-	-
11	VP test	+	+	-	-	-	-	-	-	-	-
12	Methyl red	-	-	-	-	+	+	-	+	+	+
13	Indole	-	+	-	-	-	-	-	-	-	-
14	Melanote	-	-	-	-	-	-	-	-	-	+

ONPG-O-Nitrophenyl-β-D-galactopyranoside : **VP-** Voges Proskauer : **H_2S-** Hydrogen Sulphide

4.12 IDENTIFICATION OF BACTERIA FROM AMF

The bacteria from five isolates were identified using molecular technique based on the 16S rDNA gene amplification using polymerase chain reaction (PCR) followed by sequencing of the amplified gene was presented in Fig.14. The primers binds to respective base positions of the 16S rDNA genes of *Streptococcus sp.* with 99 % similarity, *Bacillus* sp with 85 % similarity and *Brevundimonas vesicularis* with 99 % similarity.

Fig14.Agarose gel electrophoresis picture obtained after 16SrDNA amplification

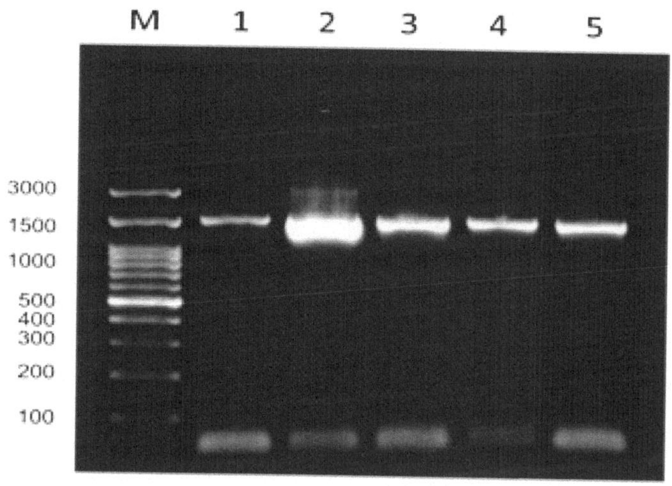

M – Marker ;**1**-R1 ; **2**-R2 ; **3**-R5 ; **4**-P2 ;**5**-P4

Table -17Sequence length and closest phylogenetic affiliation of 16s r DNA band

Sample Code	Sequence Length (bp)	Phylogenetic Relationship Species	Raceways	Lined Ponds	Similarity (%)
R1.	352	*Bacillus sp.*	+	-	85
R2.	331	*Staphylococus captis*	+	-	99
P3.	350	*Brevundimonas vesicularis*	-	+	99

[(+) presence and (-) absence of organism]

4.13 TECHNICAL FEASIBILITY AND ECONOMIC VIABILITY OF RACE-WAYS AND LINED POND SYSTEMS :

4.13.1 NURSERY RACEWAY FOR REARING *Litopenaeus vannamei*

Investment costs:

Table.18 a.Investment cost for AMF in indoor raceway culture:

S.No.	Items	Cost (Rs.in Lakhs)
1.	Raceway tanks ($50m^3$)- (2 Nos.)	1.00
2.	Green house facility	0.75

3.	Blowers (5HP & 10 HP) -2 Nos.	1.20
4.	Generator (15 KVA)	1.00
5.	Rapid sand Filter (20000 LPH)-1 No.	0.40
8.	Borewell,pump and airline	0.50
10.	Miscellaneous	0.25
	Total	5.10

Table.18 b.Variable costs: (To raise one nursery crop in raceways)

S.No.	Items	Cost (Rs.in '000)
1.	Fertilizers and Carbon sources	4000.00
3.	Shrimp seeds (6000 PL./m^3)	60,000.00
4.	Feed (30 % protein)	14185.00
5.	Power	25000.00
6.	Fuel (Gen set)	2000.00
7.	Labour	6000.00
8.	Interest on investment cost (appointed a crop)	10200.00
9.	Miscellaneous	2000.00
	Total	1,23,385.00

Table.19 ECONOMIC FEASIBILITY ANALYSIS

S. No.	Items	
1.	Yield (91 % survival)	2,18,400 Nos.
2.	Possible crops per year from raceways	5
3.	DOC	50 days
4.	Total no.of seeds to be produced	1.09 million
5.	Cost of production of 1 juvenile	Rs.1.30
6.	Total expected returns	Rs.14.01 lakhs
	NET RETURNS PER ANNUM	Rs.11.26 -Rs.14.01 =Rs.2.94 lakhs
	BENEFIT COST RATIO	1.24

4.13.2 NURSERY Lined pond FOR REARING FOR REARING *Litopenaeus vannamei* in lined ponds

Table. 20.a. Investment cost for AMF driven lined ponds:

S.No.	Items	Cost (Rs.in Lakhs)
1.	LDPE lined LWE growout ponds (100m^3) -2 no.s	1.80
2.	Aspirator and accessories (4 Nos.)	2.00
3.	Pump (5HP) and water line	0.75
4.	Miscellaneous	0.30
	Total	Rs.4.85 lakhs

Table.20.b. Variable cost for AMF and non-AMF driven lined ponds:

S.No.	Items	Cost (Rs.in Lakhs)
1.	Fertilizers and Carbon sources	3000.00
3.	Shrimp seeds (2300 PL./m^3)	30,000.00
4.	Feed (30 % protein)	11044.00
5.	Power	20000.00
6.	Fuel (Gen set)	2000.00
7.	Labour	6000.00
8.	Interest on investment cost (appointed a crop)	26000.00
9.	Miscellaneous	2000.00
	Total	1,00,044.00

Table.21. Economic Feasibility Analysis for AMF driven nursery culture in outdoor lined ponds:

S. No.	Items	
1.	Yield (86 % survival)	1,54,00 Nos.
2.	Possible crops per year from lined ponds	5
3.	DOC	50 days

4.	Total no.of seeds to be produced	7.7L
5.	Cost of production of 1 juvenile	Rs.1.50
6.	Total expected returns (@ Rs.1.5 per juvenile)	Rs.11.55 Lakhs
	NET RETURNS PER ANNUM	Rs.9.85.00 -Rs.11.55 =Rs.1.7 lakhs
	BENEFIT COST RATIO	1.17

The benefit cost ratio of the nursery culture in raceways and lined ponds was above 1.00 and this indicates the two systems are economically feasible. However the benefit cost ratio of nursery raceways is higher compared to lined pond culture.

5

CHAPTER

SUMMARY OF THE FINDINGS

Super-intensive shrimp culture technology allows the use of smaller systems with consistent production, versatile location options and high product quality for unique product marketing opportunities. The AMF technology which involves the nutrient recycling for the effective utilization of feed in concomitant with the natural bioremediation will serve the aquaculture at its fullest. AMF is not only a science fiction rather it improves the fish production with minimum water for making the business aquaculture more successful. This research also brings an insight to the farmers for taking the aquaculture to the controlled system in the indoor conditions. There is a great lack of biotechnologies for the development of species in AMF, which highlights the need to establish coordination mechanisms for conservation and for the management of resources. Results are highly encouraging and demonstrate a substantial potential for AMF improvement in aquatic productions, in particular for species such as growth rate and resistance to diseases. This would gather current knowledge and potential developments and allow appropriate and optimal planning of research required for the development of Indian aquaculture in the short, medium and long term

5.1 WATER QUALITY PARAMETERS

The salinity was maintained in about 7 ppt in raceways and lined ponds which directly influences the positive growth on shrimps, as Bray *et al* .,1994 showed that 5 to 15 ppt salinity produced significantly greater final weights than other levels tested and the hyper-saline treatment (49ppt) showed significantly lesser growth of *L.vannamei*. The highest Ammonia-Nitrogen (NH_3-N) concentration recorded in our study was 0.68 mg L_1 (at pH 7.3, temp. 29.2 C, and salinity 7 ppt) in indoor raceways and 1.3 mg L_1 (at pH 7.5, temp. 32.2 C, and salinity 7 ppt) in outdoor lined ponds. It was significantly higher in outdoor ponds due to decreased bacterial biomass in the system. Furthermore, we did not observe any dead PL during periods of high ammonia concentration it is probable that TAN did not affect shrimp survival in our trial. It is interesting to note that the patterns of nitrite concentration were different for both the system. Although the increase in nitrite concentrations started after the third week in outdoor lined ponds ,the maximum nitrite-N levels in the outdoor lined ponds were higher than the indoor raceways (1.71 vs. 0.15 mg NO_2-N L^1). This higher concentration was expected since the development of nitrifiers as well as the heterotrophs was slow in the outdoor lined ponds than indoor raceways. The low concentrations of nitrite in the indoor raceways reveals the complete oxidation of ammonia to nitrate and inorganic nitrogen conversion into bacterial biomass (Cohen *et al.*,2005, Ebeling and Timmons, 2007; Schneider *et al.*, 2007).

Nitrate concentrations averaged 24.78 mg L^1 and 98.28mg L^1 for the raceways and lined ponds, respectively, with final concentrations increasing to 45 mg L^1 and 198 mg L^1 at the end of trial (Table 9). In our study exposure to a maximum NO_3 concentration of 257 mg L^1 or mean of 216.28 mg NO_3-N L^1 had no adverse effect on shrimp and did not affect the growth and survival in indoor raceways but affected the survival in outdoor lined ponds at lower salinities (Kuhn *et al.*,2010) as it was relatively lower. In the present study, low NH_3-N and NO_2 N concentrations were observed in both systems below the safe level as cited by Kuhn *et al.*,2010 and their dynamics didn't seem to be affected by the carbon supplementation (Xu and Pan, 2012).The rapid assimilation of ammonia after the application of MF-CEED as a carbon source was experienced at faster pace in indoor raceways than lined ponds due to increased heterotrophic population. And this study confirms the influence of carbon source in the succession of heterotrophic bacteria over microalgae(Ju *et al.*,2008).

TSS,TDS, VSS and turbidity changes followed the same increasing trend as nitrate, reaching levels higher than 237 mg L^1, 15.6 mg L^1, and 131 mg L^1, 79.2 NTU respectively at the end of the study in indoor culture. These values in the raceways were significantly higher than the lined ponds(Table 9).The less floc volume and TDS in ponds indicate the reduced population of bacteria, cyanobacteria, rotifers and nematodes (Ray *et al.*,2010b) in lined ponds. The higher turbidity in raceways may be due to the presence of total suspended and dissolved solids (Ray *et al.*,2014). However, the levels of TSS and TDS during the experiment showed direct correlation with regard to the carbon supplementation and disagrees with the findings of Wu-Jie Xua *et al.*,2016.The phosphate levels in the outdoor lined pond treatment was significantly higher than the indoor raceways. The slight decrease of phosphate concentration in the raceways dominated by heterotrophic bacteria may have been due to assimilation of phosphate by heterotrophic bacteria (Longnecker *et al.*, 2010 and Zubkov *et al.*, 2007).

The five-day carbonaceous Biochemical Oxygen Demand (BOD_5) ranged from minima of 33.21 and 51.44 to maxima of 145.20 mg L^1 and 121.40 mg L^1 in the indoor raceways and outdoor lined ponds respectively. Higher BOD_5 indicates that microbial respiration and oxidative processes consumed a greater amount of oxygen in the heterotrophic dominant systems (Azim and Little,2008 ; Browdy *et al.*, 2012).The average dissolved oxygen levels in the indoor raceways were above 5.6 mg L_1. The AM and PM maxima and minima DO concentrations reported in Table 9 suggest that DO never dropped below 3.6 mg L_1.The concentration of CO_2 was relatively higher in raceways (0.5-7.10 ppm) than lined ponds (0.9-3.10 ppm) and this may be due to the regeneration of ammonia by the dominant population of heterotrophic bacteria in the raceways(Avnimelech, 2015).

In both system the pH, alkalinity, temperature were within the acceptable range for the species (Van Wyk *et al.*,1999).All these water quality results suggests the heterotrophic bacteria have a maximum growth rate significantly higher than nitrifiers, replicates 5 times per day compared to one replication per day, and are thus able to outcompete autotrophs in indoor conditions (Ebeling *et al.*, 2006). This increased trend towards heterotrophy with increasing carbon supplementation was also noticed highly in the indoor raceways. The conversion of NO_3-N or NO_2-N to nontoxic N_2 gas by anaerobic denitrifying bacteria can be enhanced by giving the suitable conditions and these bacteria are also capable of assimilating PO_4 and they typically generate alkalinity (Hamlin *et al.*,2008, Ray *et al.*,2011).

5.2 WATER EXCHANGE RATE

The AMF system in indoor raceways saved about nearly 3times of the water needed to maintain nontoxic conditions during vannamei production in outdoor culture. Additional water to maintain AMF water quality in indoor raceways was 68m^3, while water to maintain AMF culture was 272 m^3 in outdoor ponds. Besides the reduction in water consumption, the application of AMF also brings other beneficial effects such as better water quality which, leads to the more environmentally friendly aquaculture practices and increase biosecurity which can control the transmission of the animal diseases (Tacon *et al.*,2002)

5.3 GROWTH PARAMETERS

There was a significant difference ($p<0.05$) in final mean body weight ,Protein Efficiency Ratio (PER) and Yield in the nursery culture of vannamei between raceways and lined ponds. There was no significant difference between raceways and lined ponds in Specific Growth Rate, Feed Conversion Rate and Survival. The growth of vannamei did not differ significantly between indoor raceways and outdoor lined ponds and this results supports the findings of Krummenauer *et al.* 2011.,as the growth of shrimp does not affect even at the higher stocking density (Arnold *et al.*,2006b,Wasielesky *et al.*,2013).

The survival of the vannamei in ponds was relatively higher in raceways and this may be due to the effective aeration system provided by the air blowers.The synergistic effects of the improved water quality and bacterial densities leads to the overall effective control of all parameters .As the aspirators in lined ponds could not offer the sufficient aeration and dead ends were formed in the corners of the pond. As bacteria require substantial amounts of oxygen to assimilate ammonia and nitrifying bacteria can be slow to establish, resulting in spikes of toxic ammonia and nitrite (Ray et al., 2009). The oxygen requirement in AMF technology is high to maintain the correct balance between microbial flocs and culture animal. The SGR values in our study for 50 days (7.0-7.2% increase in weight/day) was closely similar to the values reported for stocking densities of 3000-6000 animals/m^3 (8.6-9.2 % weight increase/day (Esparza-Leal *et al.*,2015) for 42 days in BFT systems. The adequate water quality concentrations allowed shedding and proper formation of the exoskeleton, promoting growth and survival in the raceways even at the higher stocking density (McGraw and Scarpa,2003).

In this study, the FCR was not significantly different between systems (p>0.05) and the addition of carbon sources (MF-CEED) increases shrimp growth and reduce FCR (Arnold *et al.*, 2006, Arnold *et al.*, 2009 and Audelo-Naranjo *et al.*, 2010). However decreased FCR was found in photo-autotrophic outdoor system than compared to heterotrophic based indoor system (Ray *et al.*,2009). This indicates the significant utilization of AMF by vannamei in both the systems. The Protein Efficiency Ratio(PER) was significantly higher in outdoor lined ponds as autotrophic dominated system are rich in protein microbial mass compared to bacteria dominated heterotrophic system and this leads to the assimilation of microbial protein by the cultured animals (Avnimelech,2015). However the efficiency of protein utilization is twice as high in AMF technology systems in both systems making it a low-cost sustainable constituent to future aquaculture development (De Schryver *et al.*, 2008). There was a significant difference in the yield between the culture system and this is attributed due to the increased stocking density in raceways.This study strongly supports the closed aquaculture system *i.e* raceways didn't inhibit the growth and survival of vannamei and differ with the findings of Kuhn *et al.*,2010.

In this study, at the higher stocking densities (6000 PL/m^3) the final mean weighty was lower in indoor raceway system than at low stocking densities (2300 PL/m^3), this indicates that *L. vannamei* growth is density-dependent during the nursery phase of production. Also, results from this study indicate that stocking density can be increased from 2300 to 6000 PL/m^3 without significantly compromising shrimp growth in AMF technology. Density-dependent growth has been reported also for *L. vannamei* (Williams *et al.*, 1996), as well as for other penaeid species (Emmerson and Andrews, 1981; Ray and Chien, 1992; Williams *et al.*, 1996; Martin *et al.*, 1998; Coman *et al.*, 2004). However, if good water quality and food availability are maintained, the negative relationship between stocking density and shrimp growth may be uncoupled up to a certain threshold density, beyond which shrimp behavior may suppress growth. These data are comparable to those reported by other authors, who indicate that at stocking densities > 3300 < 17,600 orgs/m^3 the *L. vannamei* postlarvae may grow between 0.23 and 1.12 g at 35–62 days of culture from PL 8-10 with survivals between 84 and 97.5% (Cohen *et al.*, 2005; Wasielesky *et al.*, 2013; Correia *et al.*, 2014)

5.4 FEED ECONOMICS

Based on the theoretical and practical calculation, the application of AMF is believed to reduce the feed cost. This is in agreement to other studies (Avnimelech.,2007 and Hari *et al.*,2004 and 2006).This study also proves the reduction on feed cost at high stocking density in raceways thereby sustainable intensification with good returns can be ensured in shrimp nursery culture .

5.5 PROXIMATE COMPOSITION OF AMF

In the proximate composition, crude protein, crude lipid and total ash were significantly different ($p<0.05$),indicating that the AMF quality was not only related to the quality of the feeds and carbon source but also depends on the environmental conditions of the culture systems. Proximate analysis of the AMF in the present study agrees with the findings of Ballester *et al.*,2010 who reported 30.4% crude protein (CP) with wheat flour and molasses as carbohydrate sources. The crude protein content of AMF collected from two systems were within the range of what have been previously studied (Azim & Little 2008; De Schryver & Verstraete 2009; Crab *et al.* 2010; Ekasari *et al.* 2010;Ju et al.,2008a; Maica et al.,2012;Emerciano et al.,2012b). The lower crude protein level was noticed from the earlier findings (Crab *et al.*, 2010 and Ekasari *et al.*, 2010) in the harvested floc collected from indoor raceways. This may be due to differences in bacteria taking part in floc formation (Rittmann and McCarty, 2001).

Ju *et al.* (2008a) reported that chlorophyll-dominated biofloc contained higher crude protein content (42%) than flocs dominated by diatoms (26–34%) and bacteria (38%). This further suggests that the microbiota that constitutes the floc is likely to affect the protein content of the bioflocs (Ju *et al.*,2008) .The crude lipid was significantly higher in indoor raceway culture than outdoor lined ponds and this may be attributed by the presence of high diatom density as it contains the lipid upto 25% (Shifrin & Chisholm.,1981).High level of ash in floc (45%) was closely similar to the findings of De Schryver and Verstraete .,2009 and this may be due to the increased solid concentration and type of carbon source used. The higher level of gross energy has been recorded by using MF-CEED as carbon source, the reason for this is not clearly known but this may be directly related with the utilization of floc by the shrimps in *in situ* culture as reported by Hargreaves, 2006 and Azim *et al.*, 2008. The MF-CEED as a carbon source enriched with rice flour was better therefore, it can be speculated that the AMF produced from MF-CEED provided better nutrition, contributing to shrimp performance.

5.6 FATTYACID PROFILE OF AMF

Out of 11 fatty acids detected in Aerobic Microbial Floc, palmitic acid, linoleic acid and oleic acid were the dominant fatty acids, and omega 3 fatty acids were detected in optimum level in both the systems (Crab *et al.*,2010).However, linolenic acid C18:3 (n-3) from AMF composition in outdoor culture was significantly higher and closely similar to the findings Ekasari *et al.*,2010 who reported 11.95 % in biofloc collected from bioreactors using different carbon sources. This may be probably due to the positive influence of MF-CEED with the autotrophic system compared to the heterotrophic system. Comparatively lower level of PUFA in floc collected from raceways may be due to the dominance of heterotrophic bacteria compared to PUFA rich micro algal community in outdoor culture (Meyers and Latscha., 1997). This relatively increases the growth of juvenile in outdoor pond culture compared to indoor raceway culture (Mayra L. González-Félix *et al.*,2003) .The unsaturated fatty acids were relatively higher in outdoor ponds as it contains the ciliates, live flagellates and micro-algae predominantly than indoor raceways Lim *et al.*, 1997b; Glencross 2009.The biochemical composition of floc depends on the microbial composition and this is greatly influenced by the light and the type of culture systems. Hence, lower dominance of autotrophic community in the present study might be due to lack of direct sunlight in the floc indoor raceway facility Ju *et al.*,2008a and dominance of algal communities over bacterial biomass in flocs collected from outdoor lined pond shrimp culture. Current method of AMF production using nitrogen fertilizers and MF-CEED as carbon source in indoor raceways at higher stocking density is feasible and easier compared to bioreactors.

5.7 PLANKTON COMPOSITION OF AMF

Shrimp, post larvae and juveniles are able to consume a little micro algae, cyanobacteria and other micro organisms directly from the water column (Kent *et al.*, 2011).Nematodes, ciliates and copepods were present in the larger population and were seen almost exclusively grazing within the floc. The phytoplankton was low in the indoor raceways due to the limited light. It has been demonstrated that plankton could serve as a supplemental food source for *Litopenaeus vannamei* (Anderson *et al.*,1987;Chen *et al.*,1992) and therefore could increase the conversion efficiency of microbial protein into *L.vannamei* protein.

An important aspect to emphasize is the seasonal variation in the microbial communities mainly affected by ecological succession (predation-competition phenomenon) especially when the cultures are performed outdoor (Vezzulli *et al.*,2002). In outdoor cultures with enough light and natural photoperiod, the diatoms were the dominant community (Godoy *et al.*,2011), while under indoor conditions with low light or lack of light ,the heterotrophic bacteria may dominate (Becerra-Dorame *et al.*,2011).The ciliates, a protozoan community was dominated at low salinities in the outdoor culture (Maica *et al.*,2012). The diatom population was found highly in both the systems at the later part of the culture (Ju *et al.*, 2008). It has a major role in maintaining water quality through the process of photosynthesis (Hargreaves, 2006). Relatively low abundance of dinoflagellates throughout the study in the indoor and outdoor systems was encountered.

5.8 BACTERIAL COMPOSITION OF AMF

High survival recorded in indoor raceways was due to the dominance of heterotrophic bacteria under lower light level and intensity can affect bacterial community structure, which has a significant effect on the survival and production of the cultured species (Lopez-Tarin., 2011). The heterotrophic count in indoor raceway was far higher than by the earlier findings (Becerra-Dorame *et al.*, 2011) in *L. vannamei* nursery. MF-CEED shows high 2×10^7 heterotrophic bacterial count in indoor culture and clearly indicates that MF-CEED supports the growth of fungi, yeast and bacteria. The range of bacteria 10^7 to 10^8 cells mL^{-1} from the present study in indoor raceways and outdoor ponds were in compliance with the findings of Avnimelech., 2012, Otoshi *et al.*, 2006 and Bauman and Pearson., 2003. Improved survival found in indoor raceways was also exhibited by the stable bacterial community in water (Thompson et al., 1999). In 'young' floc for instance, heterotrophic bacteria are mainly abundant, whereas in 'old' floc, fungi took its prominence (Kuhn & Lawrence, 2010).The presence of Bacillus predominantly in the indoor raceways had the lower vibrio load when compared to the outdoor pond culture and inhibited the proliferation of pathogens (Crab.,2010). The *Bacillus sp.*, identified from this study has the tendency to grow efficiently with low-cost carbon and nitrogen sources, because its enzymes are very efficient breaking down a great variety of proteins, carbohydrates and lipids from animal and vegetable origin, into their constituent units (Sonnenschein *et al.*,1993, Arellano & Olmos., 2002). *Brevundimonas vesicularis*, isolated and sequenced from outdoor culture can accumulate the poly-beta hydroxybutylate from effective carbon sources(Jang *et al.*,2010). *Staphylococcus capitis* should be considered as opportunistic bacteria with potential to become pathogenic for fish under stressful conditions. It is not possible to eliminate the bacteria from the fish or from the environment (Varvarigos., 2001).

Techno-economic viability of AMF driven Nursery culture in indoor raceways and outdoor lined ponds

As nursery rearing of *Litopenaeus vannamei* in raceways and lined ponds sounds economically feasible in both the culture system. However, the higher benefit cost ratio in raceways clearly indicates the rearing of vannamei at higher stocking density will be more profitable compared to outdoor lined pond culture. It is possible to raise the production level to many folds as this system ensures high survival. Further, the quality of culture practice adopted in such systems are at its high level. The hi-tech practice, using bioremediation techniques in such system have proved to be the successful formula for enhanced shrimp production. Its technically feasible, economically viable and it is also eco-friendly for adoption(Felix.S.2014).

6 CHAPTER

SUMMARY

A 50 days nursery trial was conducted on *Litopenaeus vannamei*, stocked at 6000 PL /m^3 in 50 m^3 raceways and 2300 PL /m^3 in 100 m^3 lined ponds in duplicates. Microbial Floc-Carbohydrate Enriched Distillery Effluent (MF-CEED), obtained as distillery spent wash from the distilleries was used as the carbon source to maintain a C:N ratio at 15:1. The commercial feed containing 30% crude protein (CP), 7% lipid and 4% fibre were fed to the animals. Nitrate-N was significantly greater in the outdoor lined ponds, accumulating to a peak mean concentration of 216 mg/L NO_3-N whereas, it was below 24 mg/L NO_3-N in the indoor raceways. The raceways showed significantly higher 5-day biochemical oxygen demand (BOD_5) and total dissolved solids concentration due to increased bacterial biomass. The final mean body weight and protein efficiency ratio were significantly higher in outdoor lined ponds whereas survival and yield were significantly higher in indoor raceways. The Aerobic Microbial Floc (AMF) driven shrimp nursery culture in indoor raceways saved about nearly 3 times of the water needed to maintain nontoxic conditions in outdoor culture. The feed economic analysis indicates that 8 % gain in terms of feed costs kg^{-1} shrimp juvenile live weight produced in outdoor ponds compared to indoor raceway culture .The biochemical composition of floc obtained from indoor raceways showed rich crude lipid and total ash whereas crude protein was significantly higher in outdoor lined ponds. The unsaturated fatty acids composition of AMF was significantly higher in outdoor culture. This study demonstrates that MF-CEED has influenced positively in the biochemical composition of floc in the outdoor culture compared to indoor raceways culture. The number of heterotrophic bacteria in the water was greater in the Indoor raceways.16SrDNA regions of some dominant species were sequenced and identified. These results indicate that differences in the culture environment on the application of MF-CEED can lead to substantial disparity in system function and shrimp production. Thus, the nursery culture of *Litopenaeus vannamei* stocked at 6000PL/m^3 in indoor raceway system with good survival (91.1%), yield (4.92 kg/m^3) and low FCR (1.1) ensures the predominant growth of heterotrophic bacteria with effective water quality compared to the outdoor lined ponds for the profitable shrimp culture in the AMF technology.

INDEX

9 789354 614170

Printed at Repro India Ltd.